Volcanoes in eighteenth-century Europe

An essay in environmental humanities

Oxford University Studies in the Enlightenment – formerly *Studies on Voltaire and the Eighteenth Century* (*SVEC*), is dedicated to eighteenth-century research.

http://www.voltaire.ox.ac.uk/our-books/
for-authors-studies-enlightenment

Volcanoes in eighteenth-century Europe

An Essay in Environmental Humanities

DAVID McCALLAM

Published by Liverpool University Press on behalf of

ISBN 978 1 78694 229 6
Oxford University Studies in the Enlightenment 2019:07
ISSN 0435-2866

Voltaire Foundation
99 Banbury Road
Oxford OX2 6JX, UK
www.voltaire.ox.ac.uk

A catalogue record for this book is available from the British Library

The correct style for citing this book is
David McCallam,
Volcanoes in eighteenth-century Europe: an essay in environmental humanities
Oxford University Studies in the Enlightenment
(Liverpool, Liverpool University Press, 2019)

Cover illustration: Carlo Bonavia, *Veduta del Vesuvio in eruzione dalla lanterna del molo* (1757). Beaulieu, National Motor Museum.
By kind permission of Lord Montagu of Beaulieu.

Printed and bound by TJ International Ltd, Padstow, Cornwall, PL28 8RW

Oxford University Studies in the Enlightenment

Volcanoes in eighteenth-century Europe

An essay in environmental humanities

This study explores the explosive history of volcanoes and volcanic thought in eighteenth-century Europe, arguing that the topic of the volcano informed almost all areas of human enquiry and endeavour at the time.

Encountered on the Grand Tour, sought out by scientific explorers or endured by local populations in southern Italy and Iceland, erupting volcanoes were a physical reality for many Europeans in the eighteenth century. For many others, they represented the very image of overwhelming natural power, whether this was ultimately attributed to spiritual or material causes. As such, the volcano proved an effective and versatile 'tool for thinking' in a century which ushered in modernity on several fronts: continental tourism, new earth sciences, the sublime and picturesque in art, industrial and political revolution, the conception of the modern nation-state, and early intimations of environmental and climate change. But the volcano also gives us, in the twenty-first century, a privileged site (as both topography and topos) at which we can reconnect disparate and divided fields of research across the sciences and the humanities.

Drawing on a rich variety of multilingual primary sources and the latest critical thinking, this study combines material and symbolic readings of eighteenth-century volcanism, constantly shifting frameworks, so as to consider this topical object through different disciplinary perspectives. The volcano is clearly transnational; this research also demonstrates how it is fundamentally transdisciplinary.

Contents

List of illustrations

Acknowledgements

As befits its subject matter, this book has evolved at an almost tectonic pace. I have worked on this project on and off for almost fifteen years. Over this time I have accumulated a lot of intellectual and collegial debts of gratitude, too many in fact to do justice to them all here (but you know who you are). I would nonetheless like to express my thanks to the International Society for Eighteenth-Century Studies and the British Society for Eighteenth-Century Studies for allowing me public platforms from which to air the first iterations of some of these ideas. I am also very grateful to my colleagues and friends in the Department of French, and more generally the Faculty of Arts, at the University of Sheffield for having enquired so long and with such good humour after 'the volcano book'. David Pyle and latterly John Brewer have both provided insightful advice and telling information regarding eighteenth-century volcanism. Louise Lyle remains the best guide to Paris for the itinerant researcher. And I would also like to thank those indomitable mountain spirits – Anton Hookes, Matt Ryder, Natalie Grove, not forgetting Dylan Ryder and Manon Ryder – with whom I have discovered the manifold delights of volcanic Sicily and southern Italy. I owe more than anything to my father, mother and brother, to my wonderful daughter Eve and son Isaac who have undoubtedly lived longest with this book. And above all, I could not have completed this research project without the unfailing love, encouragement and life-affirming presence of my wife Renata Schellenberg – my very own force of nature.

Introduction

This study traces the volcano's physical and imaginative eruptions into eighteenth-century European consciousness. One of its principal aims is to demonstrate how the topic of the volcano – as both physical site (topography) and cultural cipher (topos) – informs almost all areas of human enquiry and endeavour across the continent at the time.

In effect, Europe contains a remarkably diverse range of volcanic terrain. From Krafla in the icy north to Santorini in the southern Aegean, violently active volcanoes continue to burst sporadically into fiery life. Sulphur fields smoulder at Pietramala north of Florence and, more famously, in the Campi Flegrei west of Naples. There are volcanic 'hotspots' still glowing beneath western Germany at Eifel and at Garrotxa in Catalonia. The rolling massif of central France is shaped by volcanism, as are the prismatic basalt shorelines of Northern Ireland and the Scottish isles.[1] Most of these volcanic phenomena were first understood as such, explored, mapped, discussed and exploited in the eighteenth century. Hence the European landmass and its islands provide a rich and revealing range of materials for this study and circumscribe its geographical limits. In other words, a conscious choice has been made not to extend its scope to encompass other volcano-rich areas of the world under exploration at the time, such as the eruptive mountain chains of South and Central America or the active volcanic islands of Indonesia, the Philippines or Japan.

Volcanism was recognised as a pan-European phenomenon in the eighteenth century largely as a result of newly formed intellectual networks across the continent. These networks drew specifically on the observations of enlightened travellers and writers roaming over

1. See Dougal Jerram, Alwyn Scarth and Jean-Claude Tanguy, *Volcanoes of Europe*, 2nd edn (Edinburgh, 2017).

jagged, charred lava fields from Iceland to Sicily, enthusiastically sharing their findings with each other in print and in the public fora of academies, societies and public lectures. Whether expert or amateur in character, local fieldwork often sparked international debate. Volcanism thus informed exchanges that offered alternative continent-wide forms of knowledge and authority, challenging more traditional views propagated by the conservative European powerhouses of the Catholic Church and the dynastic royal courts. A keen interest in volcanoes grew particularly strongly in the relative peace in Europe between 1763 and 1792 – precisely the period in which Vesuvius and Etna were almost constantly rumbling and spurting lava and flames. The fascination with all things volcanic was fostered by the voluble cosmopolitanism of the Grand Tour – the cultured travels of young, usually male Britons, French and northern Europeans to the artistically rich cities of northern Italy and Rome – which often culminated in Naples or occasionally extended into Sicily. Hence it is important to state that a significant number of educated Europeans witnessed first-hand a volcano in one eruptive phase or other. The experience proved, on the whole, to be terrifying, exhilarating, exhausting, unforgettable. It affected its beholders on a personal and a collective level; it was at once particular to the individual's sensibilities and universal in its intimations of human frailty, transience and mortality. Encountering the volcano was thus local and global in significance and was only to be appropriated nationally, in this age of nascent nation-states, through the volcanic upheavals of the French Revolution. The volcano was a shared European experience in at least two ways: it was increasingly identified as a continent-wide agent of geological and geographical formation, and it conveyed a continent-wide expression of the eighteenth century's unique intellectual effervescence.

Tzvetan Todorov has claimed that the complementary and contradictory movements of the so-called Enlightenment arose specifically in eighteenth-century Europe because European states exercised competing forms of political and cultural autonomy. Geographically concentrated, this pluralism bred a critical spirit keen to distinguish between the divergent political and cultural practices that it found across neighbouring territories.[2] In an essay of 1742 David Hume compared the special intensity of intellectual activity in eighteenth-century Europe to that of fifth-century-BCE Greece. His comparison suggested that the peculiarly rich philosophical ferment of the ancient

2. Tzvetan Todorov, *L'Esprit des Lumières* (Paris, 2006), p.122-39.

Greek peninsula, composed of independent city-states and diverse island cultures, prefigured the critical plurality driving the intellectual 'commons' of Enlightenment Europe.[3] Eighteenth-century Europe thus provides a coherent, if far from consensual, epistemological framework for approaching volcanism in all its physical and cultural manifestations. It constitutes an 'episteme', to borrow Foucault's term, affording both continuity and rupture with its 'classical' seventeenth-century foundations.[4] As such, its understanding of volcanoes at once consolidates and challenges the preceding century's prevailing conceptions. Indeed, eighteenth-century European thought often found in the volcano an apt figure of nature's own conflicting imperatives of timeless continuity and spectacular rupture, imperatives which also characterised the intellectual ambitions of the age.

For contemporary Europeans this meant that the volcano appealed equally to an enduring classical mindset and to emerging modern thinking, to both tradition and innovation. For example, the image of Etna as home to the Cyclops' terrible forges in Virgil's *Georgics* spoke at once to a knowledge of ancient Greek and Roman myth as well as a desire to harness volcanic energy to evolving heavy industry. The volcanic was both primitive and progressive. On the one hand, it stood for the elemental force of fire in nature – as the hurricane stood for that of air, the tidal wave for that of water, and the earthquake for that of earth. On the other, the volcano was integrated into complex new Earth histories, for example, in the way in which Georges-Louis Leclerc de Buffon deployed it as a dynamic component in his shifting understanding of the formation of the globe. In both instances, however, the volcano represented a timely geocentric correction to contemporary anthropocentric hubris. This also chimes with recent eco-critical research, insofar as it engages with the controversial idea that the planet has entered the new geological epoch of the 'Anthropocene', defined by humankind's irreversible impact on its natural environment.[5] While certain critics claim that the Anthropocene began in late-eighteenth-century Europe, it is worth noting that

3. David Hume, 'Of the rise and progress of the arts and sciences', in *Essays moral and political*, 2 vols (Edinburgh, A. Kincaid, 1742), vol.2, p.53-100 (71-73). Also cited in Todorov, *L'Esprit des Lumières*, p.130-33.
4. See Michel Foucault, *Les Mots et les choses: une archéologie des sciences humaines* (Paris, 1966), p.13-14.
5. The standard reference here is to Paul J. Crutzen and Eugene F. Stoermer, 'The "Anthropocene"', *IGBP newsletter* 41 (May 2000), p.17-18.

enlightened Europeans were already interrogating the power relations between humanity and geology, and volcanism played no small part in these interrogations.[6]

It remains less clear at what precise historical point the particular episteme of Enlightenment Europe was superseded by another order of thought and practice. It could be argued that many aspects of eighteenth-century thinking persist and inform every subsequent age up to the present. But it also appears certain that the economic dominance of the middle classes, the emergence of various Romantic movements in art and literature, and the conservative post-revolutionary political order in Europe mark a tangible epistemic shift from what preceded them. As far as the volcano is concerned, this shift is experienced mainly as a form of parody or exaggeration. The Grand Tour is commercialised; its quest for refined feelings of dread and delight at the bubbling craters of Vesuvius and Etna is vulgarised; its picturesque art is reproduced cheaply as souvenirs. The raw excess of eruptive power depicted in Joseph Wright of Derby's more sublime paintings of Vesuvius in 1774 becomes a swirling fiery trope in the vast historical and biblical tableaux of John Martin or Joseph Mallord William Turner.[7] We move from a terrifying excess of matter to a gaudy excess of feeling. Volcanic sublimity gives way to volcanic melodrama. This is illustrated by three examples of garish volcanic works of the early nineteenth century: Giovanni Pacini's two-act operetta *L'Ultimo giorno di Pompei* (1825), Karl Briullov's huge canvas *The Last day of Pompeii* (1830-1833) and Edward Bulwer-Lytton's immensely popular potboiler *The Last days of Pompeii* (1834), allegedly inspired by the Briullov painting. Crucially, all three reduce the volcano to a narrative of the cataclysmic events of 79 CE. As John Brewer has suggested, this signifies an important ideological shift in early-nineteenth-century Europe.[8] After the conclusion of the Treaty

6. See Will Steffan *et al.*, 'The Anthropocene: conceptual and historical perspectives', *Philosophical transactions of the Royal Society* 369 (2011), p.842-67. For a deconstruction of the homogenising notion of 'humanity' as a global geological and climatic agent, see Karsten A. Schulz, 'Decolonizing the Anthropocene: the mytho-politics of human mastery', in *Critical epistemologies of global politics*, ed. Marc Woons and Sebastian Weier (Bristol, 2017), p.46-62.
7. See James Hamilton, 'Les volcans dans l'art et la littérature britannique au XIXe siècle: un bref survol', in *Le Vésuve en éruption: savoirs, représentations, pratiques*, ed. Emilie Beck Saiello and Dominique Bertrand (Clermont-Ferrand, 2013), p.129-39.
8. See John Brewer, 'Sublime tourism, enlightened science and counter-revolution:

of Vienna in 1815, the iconic eruption of Vesuvius in the first century CE is recuperated by the forces of political reaction and the Catholic Church to overwrite neo-classical and revolutionary conceptions of volcanism prevalent in the preceding century. We return to a volcanic inferno that blasted the sinful cities of the pagan Roman empire (read: deist cosmopolitan Europe and especially revolutionary-imperial France) the better to resurrect them 1700 years later as proof of God's wrath and the need for renewed piety and submission to the powers-that-be. In the light of these significant changes in European socio-economics, aesthetics and politics in the first three decades of the nineteenth century, this study limits its epistemological study of the volcano to the period 1700-1800.

What is designated by the term 'volcano' in this time frame poses much less of a problem, since the principal European languages used in this research share an etymology of the term. In the first instance, it derives from the Latin *volcanus* or *vulcanus*, originally a name of volcanic sites associated with the lame and cuckolded god of fire, Hephaestus in Greek and Vulcan in Roman mythology. It referred specifically to the Aeolian Islands north of Sicily, which include the ever-active Stromboli and the southernmost island-crater which is now called Vulcano. The term was also applied to Etna, which was known until at least the seventh century CE as 'Vulcanus mons'. In Italian the word became *vulcano* or *volcano*, although it was still related topographically to Etna and the Aeolian archipelago. It was principally the Spanish colonists, returning from the Americas from 1524 onwards with reports of terrifying 'montañas de fuego' or mountains of fire, who introduced the term 'volcán' as a common noun in Europe.[9] Thus 'volcano' has both an empirical and a colonial signification to complement and extend its classical Latin roots. By the eighteenth century, therefore, the noun had been in circulation for at least two centuries. What the so-called age of Enlightenment would popularise in turn was the adjective 'volcanic'.[10] Again deriving principally from

three versions of Vesuvius and Pompeii in the age of Romanticism', research presentation given at the University of Durham, 25 February 2016, https://www.dur.ac.uk/resources/ias/audio/Brewer.mp3 (last accessed 3 December 2018).

9. See 'Volcano, n.', in *Oxford English dictionary online* (3rd edn, March 2013), http://www.oed.com.sheffield.idm.oclc.org/view/Entry/224428?isAdvanced=false&result=1&rskey=IUv6pX& (last accessed 3 December 2018).

10. See 'Volcanic, adj. and n.', in *Oxford English dictionary online* (3rd edn, March 2013), http://www.oed.com.sheffield.idm.oclc.org/view/Entry/224417? (last accessed 3 December 2018).

experiences in the field, the Italian term 'volcanico' was applied to phenomena associated with Vesuvius' powerful eruption of 1754.[11] By the 1770s the adjectives 'vulkanisch' in German, 'volcanic' in English and 'volcanique' in French were also common currency, as is evidenced respectively by Johann Jakob Ferber's *Briefe aus Wälschland über natürliche Merkwürdigkeiten* (1773) and Sir William Hamilton's lavishly illustrated, bilingual *Campi Phlegraei: observations on the volcanos of the Two Sicilies* (1776).[12] In contrast, the terms 'volcanism' and 'volcanology' are nineteenth-century neologisms.

In order to capture the multifaceted character of eighteenth-century thought about volcanoes, we thus return insistently to its multilingual texts and other works. This means drawing on various English, French, Italian and German sources, and more rarely on writings in other European languages such as Czech or Spanish.[13] In keeping with the period, French predominates (this may also be informed by a certain *déformation professionnelle* on my part). The French language was, after all, the lingua franca of European cultural exchange, scientific correspondence and international diplomacy in the period. English sources are also legion; this reflects less the international standing of the language than the fact that the British, along with the French, embraced the Grand Tour most enthusiastically and travelled extensively on colonial missions often under cover of disinterested,

11. On the significant impact of Italian on the evolution of volcanological terminology, see Rosa Casapullo, 'Note sull'italiano della vulcanologia fra Seicento e Settecento', in *Napoli e il Gigante: il Vesuvio tra imagine, scrittura e memoria*, ed. Rosa Casapullo and Lorenza Gianfrancesco (Soveria Mannelli, 2014), p.13-53.
12. Johann Jakob Ferber, *Briefe aus Wälschland über natürliche Merkwürdigkeiten* (Prague, Wolfgang Gerle, 1773), p.22, 53, 59, etc.; Sir William Hamilton, *Campi Phlegraei: observations on the volcanos of the Two Sicilies*, 2 vols (Naples, n.n., 1776), vol.1, p.11, where it is spelled 'Volcanick'.
13. The main languages of Iberia are interesting insofar as they tend to subsume volcanism under seismic activity in the wake of the devastating Lisbon earthquake of November 1755. Otherwise, the Spanish accounts are displaced either historically (relating to the 1631 and 1694 eruptions of Vesuvius when Naples was ruled by the Spanish Habsburgs) or geographically (accounts of Central and South American volcanoes or José Torrubia's writing on the volcanic Philippines). Hence they do not figure prominently in this study. For the last two examples cited, see respectively Laura Rodríguez Fernández, 'Il barocco spagnolo di fronte al Vesuvio: Juan de Quiñones, *El monte Vesuvio, aora la montaña de Soma* (Juan González, Madrid, 1632)', in *Napoli e il Gigante*, ed. R. Casapullo and L. Gianfrancesco, p.111-36; and José Torrubia, *Disertación histórica política, y en mucha parte geográfica de la islas Filipinas* (Madrid, n.n., 1753).

scientific quests. This is very much borne out by the final chapter of this book on the exploration of eighteenth-century Iceland. The Scandinavian languages proved hardest to access in the original, especially Icelandic. Hence these texts are often cited via their eighteenth-century English, French or German translations, largely as an educated contemporary European readership would have consulted them. Given this multilingual pool of materials, the practice adopted here is to provide all titles in the original language with footnoted references to the foreign-language source-text, but to translate all quotations into English. These are my translations, unless otherwise stated. The original foreign-language text is not reproduced in footnotes or endnotes, as this would generate an unnecessarily large and distracting amount of annotations. As for the bewildering array of weights and measures used in our eighteenth-century sources, their original terms are retained when the units have a current standard equivalent, for example, miles, feet, inches, pounds or tons. But when obsolete weights and measures are used, such as the 'Danish mile' or the 'toise', they are translated into comparable imperial or metric units.

The book is organised into six chapters based on six loosely intersecting fields of cultural activity in the period. This could be described somewhat reductively as successive studies on the cultural volcano, the scientific volcano, the aesthetic volcano, the philosophical volcano, the political volcano and the ecological volcano (the last being a case study of Icelandic volcanoes in the eighteenth century). Chapter 1, 'From *locus classicus* to cosmopolitan picnic site', thus examines how the volcano informed neo-classical sensibilities in the eighteenth century, especially after the discovery of the ruins of Herculaneum (1738) and Pompeii (1748), and how Vesuvius in particular became a key site of cosmopolitan sociability for eighteenth-century tourists and travellers. Chapter 2, entitled 'Two modern "Plinies" and the empirical turn', contextualises the work of the two writers who most influenced scientific thinking about volcanism in the eighteenth century: Buffon's holistic theorising and Sir William Hamilton's intrepid empiricism. It demonstrates the importance of amateur fieldwork on volcanoes and across volcanic terrain in the development of European schools of early geological thought. Chapter 3, 'On the volcanic sublime, its art and artifice', explores how the volcano became a testing ground for the aesthetic theories of the 'natural sublime', as provided chiefly by Burke and Kant. It analyses the attempts of artists to capture the sublimity of active volcanoes, and the ways in which volcanic power is domesticated in pyrotechnic displays, heavy industry and tourist narratives. Chapter 4, 'More heat than light? Natural philosophies of volcanism', interrogates the volcano's role in natural philosophy from physico-theology, via theism and deism, to materialist uses of volcanism to question biblical chronologies. It charts how a fear of volcanic violence emerges as a metaphor of popular unrest, pitting Promethean visions of mastering volcanic fire against Empedoclean desires of a mystical union with it. This leads to Chapter 5, entitled 'A volcanology of Revolution 1789-1794'. It analyses the ways in which the French Revolution politicised the volcano primarily as a figure of dangerous latency and irresistible eruption. The radical faction of the 'Montagne' identified itself successfully with natural volcanic force, carrying this image into subsequent phases of the so-called reign of Terror. Abroad, the Revolution inspired widespread fears, similarly represented by volcanic imagery. Chapter 6, 'Volcanic Iceland: conquering Hekla and surviving Laki', considers the convergence of geopolitics and geophysics in eighteenth-century Iceland. Explorers on the island, often looking to 'conquer' its most famous volcano, Hekla, discovered the unique violence of Icelandic volcanism which brings together volcanoes and glaciers, fire and ice. However, it was the cataclysmic eruption of Laki in 1783 which most devastated life on the island and had a far-reaching, lethal impact on western European societies too. As such, Icelandic volcanism provides a neat case study of the wider environmental and climatic consequences of a major eruption in the period.

This study combines synthesis and analysis. The analysis concerns the primary eighteenth-century materials that deal either directly or indirectly, exclusively or occasionally, with volcanoes. These texts and artworks are analysed by close reading and extensive citation. The study is synthetic in the sense that it draws on discrete works of critical literature which touch on eighteenth-century volcanism in order to elucidate the full significance of volcanoes in Europe throughout this period. In the process, the respective claims of these secondary works are also cast in a revealing, new light. As far as the cultural import of volcanoes is concerned at the time, the key secondary sources include Chantal Grell's pioneering study of the first cohorts of tourists to explore the newly unearthed Herculaneum and Pompeii and Melissa Calaresu's enlightening analyses of eighteenth-century Neapolitan society.[14] The science of volcanoes in eighteenth-century Europe draws first and foremost on Martin Rudwick's monumental contributions to the history of early modern geology as well as more recent studies

14. See subsequent chapters for full references to the authors and works cited here.

by the likes of Sean Cocco, Corinna Guerra and Noah Heringman. Emilie Beck Saiello's superb exposition of the art of Pierre Jacques Volaire affords an indispensable point of entry for understanding the volcano in aesthetic terms in the period. This is all the more the case when it is read in conjunction with Joachim von der Thüsen's beautifully illustrated *Schönheit und Schrecken der Vulkane*. In fact, more generally, our study owes a great debt to Joachim von der Thüsen's magisterial overview of volcanoes in early modern European culture. As for contemporary philosophical interpretations of volcanism, David Constantine's excellent biography of Sir William Hamilton is highly informative and insightful, as are the works produced by Dominique Bertrand and others in the innovative *Volcaniques* series published by the Presses universitaires Blaise Pascal, appropriately located in the heart of volcanic Auvergne. To understand the political volcano of Revolution, Mary Ashburn Miller's *A Natural history of Revolution: violence and nature in the French revolutionary imagination 1789-1794* is a trove of apposite, telling detail and commentary, while Sanja Perović's study of the revolutionary calendar opens up new vistas on Sylvain Maréchal and his politicised pastoralism. For eighteenth-century Iceland, the research undertaken by Anna Agnarsdóttir is essential reading, specifically her magnificent recent edition of the documents relating to Sir Joseph Banks' North Atlantic expeditions of the late eighteenth and early nineteenth centuries. These are complemented by the extensive volcanological research that has gone into the Laki eruption over the last twenty years, led by John Grattan and Thorvaldur Thordarson among others. Finally, more general contextualisation is provided by accessible recent studies on volcanoes in social and cultural context by Davis A. Young, Alwyn Scarth and Clive Oppenheimer. Thus this book engages with important primary and secondary materials in order to expand and enhance our understanding of the interactions of volcanoes and humankind in eighteenth-century Europe. As such, it aims to make a significant contribution to contemporary Enlightenment studies.

But, as its subtitle suggests, this research also aspires to contribute to the emerging transdisciplinary field of environmental humanities.[15]

15. For an explanation of why environmental humanities might be 'transdisciplinary' rather than 'interdisciplinary', see Hubert Zapf, 'Cultural ecology, the environmental humanities, and the transdisciplinary knowledge of literature', in *Environmental humanities: voices from the Anthropocene*, ed. Serpil Oppermann and Serenella Iovino (London, 2017), p.61-79 (67).

Environmental history constitutes a privileged mode of enquiry here; but this new field also draws on eco-criticism and geo-criticism as literary interrogations of the environment (and environmental interrogations of literature), as well as environmental sciences, environmental philosophy, political ecology, ethnography, and ecological ethics and aesthetics.[16] Environmental humanities are particularly relevant to our subject because they work across the principal binaries of European or 'western' thought, especially the nature–culture divide that eighteenth-century thinkers did so much to consolidate, adumbrating in the process the later disciplinary divisions between the sciences and the humanities. Volcanism in particular reminds us that, already in the eighteenth century, the natural environment was far from being passive matter on which human thought, desires and actions were impressed. Instead, the volcano constitutes a 'landscape', at once a real and an imagined space, which disturbingly decentres anthropocentric scientific rationality and feeds into a concomitant self-critique of enlightened reason (usually exemplified by the work of Jean-Jacques Rousseau in the period). It does so in part by being the vector of alternative, equally valid modes of knowledge production (e.g. travellers' tales, local lore, political master-metaphor, iconographic object, image of humanistic or religious transcendence). But it is specifically in the unique violence and unpredictability of its eruptive events that the volcano returns agency, even creativity, to its craggy, fuming peak and so to the natural world more generally. In revisiting how volcanoes and human societies modify one another in their various interrelations and interactions in eighteenth-century Europe, a further aim of this book is thus to contribute in some modest way to what Kate Rigby has called the 'ecological enlightenment of the Enlightenment'.[17]

16. On environmental history, its evolution and relation to environmental humanities, see Grégory Quenet, *Qu'est-ce que l'histoire environnementale?* (Seyssel, 2014), especially p.272-78.
17. Kate Rigby, *Dancing with disaster: environmental histories, narratives, and ethics for perilous times* (Charlottesville, VA, 2015), p.4.

1. From *locus classicus* to cosmopolitan picnic site

When the English essayist and playwright Joseph Addison first arrived in Naples at Easter 1701, he met with a pageant of Catholic zealotry, popular superstition and venal practice in public office. He found the people lazy and pleasure-seeking, although he attributed this in part to the hot climate and easy abundance of local produce. The delightful, sweeping prospects across the Bay of Naples were much more to his liking. But above all he was struck by Vesuvius: 'there is nothing about *Naples*, nor indeed in any part of *Italy*, which deserves our Admiration so much as this Mountain.'[1] Unusually for the turn of the century, Addison made an arduous ascent of the volcano, sinking and sliding back with each step in the fine cinders on its slopes. When he finally reached the crater, the billowing steam and gases were momentarily blown aside to reveal 'a wide naked Plain, smoaking with Sulphur in several Places, and probably undermin'd by Fire [...] In the midst of this Plain stands a high Hill in the shape of a Sugar-loaf.'[2] His descriptions of the craggy desolation on the summit evoke at once uncanny confections ('sugar-loaves' of ash and 'cakes' of lava)[3] and a brooding horror of latent volcanic violence. As he peers into its treacherous, shelving crater, Addison perceives nothing sublime about Vesuvius.

Perhaps as a result of this troubling empirical experience, the volcano is translated awkwardly into the English traveller's predominantly classical vision of the region. For Addison overlays the topography of the Bay of Naples with learned allusions, concluding

1. Joseph Addison, *Remarks on several parts of Italy, etc. in the years 1701, 1702, 1703* (London, Jacob Tonson, 1705), p.202-12, 235 (on Vesuvius). Original emphasis.
2. Addison, *Remarks*, p.238.
3. Addison, *Remarks*, p.237-38 ('cakes'), 241 ('Sugar-loaf Mountain').

his observations on the area with a long citation from the first-century CE Roman poet Silius Italicus. Preparing the 'classical ground' for subsequent generations of eighteenth-century tourists, he cites the Latin verse, then translates it into English rhyming couplets. Volcanic tremors thus become the restless struggle of imprisoned Titans, while Vesuvius' smoking peaks

> (A hideous Ruin!) that with earthquakes rent
> A second Ætna to the View present.[4]

The physical volcano (Vesuvius) is only understandable as a yet more classical volcano (Etna). In this much, Addison's cultural capital of Greek and Latin learning obfuscates more than it elucidates its volcanic objects. The strenuous, sooty trek up the mountain sits a bit clumsily alongside the writer's cultured evocations of the Virgilian landscapes around Naples.

In a sense Addison sets the tone adopted by later educated visitors to Vesuvius and the Bay of Naples. His method, or 'entertainment', as he admits, is to revisit the descriptions of classical writers 'as it were, upon the spot, and to compare the Natural Face of the Country with the Landskips that the Poets have given us of it'.[5] Virgil is taken repeatedly as the primary reference, not just for poetic raptures about the green and blue vistas across the Gulf of Naples to the islands of Procida, Ischia and Capri, but also for more dismal descriptions of the steaming, lava-scarred terrain such as the nearby sulphur fields of the Campi Flegrei. The Austrian diplomat Franz von Hartig employs the same method as Addison. Visiting the Bay of Naples in 1776, he was in the habit of striking out each morning into the surrounding countryside 'with Virgil in hand'.[6] Thus, when he visited the crater lake of Averno, having passed through the acrid dust-bowl of the Campi Flegrei, his mind turned naturally to the great poet's hero Aeneas descending there into the underworld in the sixth book of the *Aeneid*. In other words, as Virgil had done before him, Hartig chooses less to see the volcanic wilderness around him than to invert its surface desolation in order to conjure up the infernal regions below ground. Steeped in classical culture, the Austrian tourist marvelled at a dank cavern on the shore of the lake, happy in the assumption

4. Addison, *Remarks*, p.250.
5. Addison, *Remarks*, 'Preface', not paginated.
6. Franz von Hartig, *Lettres sur la France, l'Angleterre, et l'Italie* (Geneva, n.n., 1785), p.235. Unless otherwise stated, all translations are my own.

that it was the famous cave of the Cumaean Sibyl, 'such as Virgil has described it'.[7]

Even those who had their doubts about the authenticity of the Roman poet's tomb – long established as a literary shrine at Posillipo some nine miles to the east of Averno – still reached for his poetry when evoking the Neapolitan landscapes. The Oxford-educated Thomas Watkins thus follows Hartig's lead during his visit to Naples in January 1788, asserting that Virgil remains 'the best and most agreeable guide [...] to this his favourite country'.[8] Watkins, whose party would later cross the Rubicon south of Ravenna 'with Lucan in our hands',[9] nonetheless redresses the balance between the erudite and the empirical, the mythological and the geological, between the powers of the classical imagination and those of deductive observation as he roamed intrepidly through this volcanic scenery. For him as for Hartig, the gloomy waters of lake Averno recalled the 'grotto of the Cumæan Sibyl'. Watkins similarly ventured into the cave system on the lake shore, although he explained how, less heroically, his party had obliged their guides to carry them on their backs through the half-flooded recesses until their way was obstructed by a rock fall.[10] More importantly, unlike his Austrian predecessor, Watkins recognises that the lake was clearly formed from the 'crater of a volcano', estimating its current circumference at 'not less than a mile and three quarters', its waters 'deep, transparent, and well tasted'.[11] Empirical observations of the volcanic terrain thus begin to overwrite the tourist's classical recollections.

Watkins could, of course, be taken as an exception to the rule, since his Grand Tour was significantly more grandiose than most. Not only did he visit Naples and climb Vesuvius twice, but he also pushed on fearlessly to Constantinople by way of Sicily and Venice before returning via Greece and the Dalmatian islands. On Sicily he attempted a perilous conquest of Etna in March, when the snows were still thick on the slopes and all but one avaricious guide refused to accompany him. Reaching the summit through a 'severe storm of sleet, with a

7. Hartig, *Lettres*, p.242-43.
8. Thomas Watkins, *Travels through Switzerland, Italy, Sicily, the Greek islands to Constantinople, through part of Greece, Ragusa and the Dalmatian isles in the years 1787, 1788, 1789*, 2 vols (London, J. Owen, 1794), vol.1, p.417.
9. Watkins, *Travels*, vol.2, p.101-102.
10. Watkins, *Travels*, vol.1, p.419-20.
11. Watkins, *Travels*, vol.1, p.421-22.

thick fog and extreme cold', he gazed in awe at the volcano's 'horrid gulph',[12] inconceivably different from the shallow basin of the crater he had seen on Vesuvius. Watkins only just completed the descent, while his long-suffering travelling companion George Pocock suffered acute frostbite on his hands and lips. It is true that learned allusions to the Cyclops Polyphemus and to the fifth-century-BCE philosopher Empedocles still colour Watkins' picaresque account of braving Etna's snow and fire.[13] But by the 1780s it might also be reasonable to suggest that such references serve principally as rhetorical ornaments to more personal, affective responses to the active volcanoes of southern Italy. Thus classical Greek and Roman culture increasingly adorns, rather than informs, the travel literature of eighteenth-century European encounters with volcanoes.

Nevertheless, some learned commonplaces proved more unavoidable and enduring than others on coming face to face with the menacingly smoking peak. One persistent reference was to Pliny the Younger's famous letters to Tacitus recounting the most iconic eruption of antiquity – Vesuvius in 79 CE. The Roman writer witnessed the cataclysm first-hand from the naval base at Misenum on the most north-westerly headland of the Bay of Naples. He claims that it began about 1pm on 24 August, a date contested by recent archaeological evidence.[14] Eighteenth-century evocations retain the famous pine-tree shape of the swelling volcanic cloud, rising nine miles into the sky; or rehearse for their readers the death of the writer's uncle, Pliny the Elder, sailing to rescue victims of the eruption but dying asphyxiated by 'dense fumes' on the beach at Stabiae.[15] They might also mention the pitch-black darkness that overwhelmed the younger Pliny's party as they fled Misenum on the second day, surrounded by the blind 'shrieks of women, the wailing of infants, and the shouting of men'.[16] Typical of the more animated renderings are Charles-Marguerite Mercier Dupaty's *Lettres sur l'Italie, en 1785*, which give a long and loose translation of Pliny the Younger's two letters, splicing their scenes together for heightened dramatic effect,

12. Watkins, *Travels*, vol.2, p.17.
13. Watkins, *Travels*, vol.2, p.16, 18.
14. See Michele Borgongino and Grete Stefani, 'Intorno alla data dell'eruzione del 79 d.C', *Rivista di studi pompeiani* 12-13 (2001-2002), p.177-215. Evidence from harvested crops, especially at Oplontis, points to a possible date up to two months after 24 August.
15. Pliny the Younger, *Letters and Panegyricus*, 2 vols (London, 1959), vol.1, p.433.
16. Pliny the Younger, *Letters*, vol.1, p.445.

presented as an edifying missive to his own son.[17] In contrast, Henry Swinburne in his *Travels in the two Sicilies* of 1790 merely bookmarks this 'woeful witness' in passing, being more concerned with describing the stunning panoramic view from the volcano's summit.[18]

Others draw on Charles Rollin's compendious *Histoire romaine* (1738-1748) in order to evoke the licentious sin-cities of ancient Campania: the baths at Baiae, the brothels of Pompeii, or more northerly Capua which had held Hannibal's troops in voluptuous thrall. The implication of these lascivious allusions appears to be that Vesuvius' eruption was nature's way to chasten the decadent and punish the vice-ridden. This would appear to be the sense of the Englishman Martin Sherlock's passing comment on visiting Naples in 1779, namely, that the volcano seemed constantly to threaten the city 'with the fate of Sodom, to consume it with fire and brimstone'.[19] Sherlock would receive a belated materialist riposte in Donatien Alphonse François de Sade's *La Nouvelle Justine* (1797), in which the libertine Cœur-de-fer reverses Sherlock's terms. Sodom was not blasted by divine fire for its excessive and unnatural sexual practices, but, being 'situated on a volcano, [Sodom and Gomorrah] were swallowed up by one of these natural phenomena whose causes are purely physical, just as the towns neighbouring Vesuvius and Etna have since been engulfed'.[20]

The disturbing discovery of Herculaneum and Pompeii

For eighteenth-century travellers the desolation of Pompeii and Herculaneum in 79 CE was not just a matter of ancient history and classical culture, it was also relatively recent news. The first inadvertent unearthing of Roman ruins dates from 1709, when a well-digger hit upon the marble paving of the buried theatre in Herculaneum. Tunnels were dug and a maze of underground galleries lit by torchlight revealed ancient statues and bright frescoes. But noxious gases stopped the visits and many tunnels were backfilled to the theatre. Further work at

17. Charles-Marguerite Mercier Dupaty, *Lettres sur l'Italie, en 1785*, 2 vols (Paris, De Senne, 1788), vol.2, p.288-300.
18. Henry Swinburne, *Travels in the Two Sicilies, by Henry Swinburne esq. in the years 1777, 1778, 1779, and 1780*, 2nd edn, 4 vols (London, n.n., 1790), vol.1, p.89.
19. Martin Sherlock, *Letters from an English traveller* (London, J. Nichols *et al.*, 1780), p.41.
20. Donatien Alphonse François de Sade, *La Nouvelle Justine, ou les Malheurs de la vertu*, 2 vols (Paris, 1987), vol.1, p.83.

the site was largely discontinued until 1738, when the king of Naples Charles III ordered more systematic excavations to be resumed. An inscription found in the theatre in December of that year confirmed that this was the ancient city of Herculaneum, buried up to a depth of twenty-three metres in compacted tephra.[21] Attention, however, switched ten years later to Pompeii when a shallower covering of four to five metres of volcanic debris was removed to reveal its treasures. Initially, many of the more impressive finds were filled in again by workers so as to be dramatically disinterred for successive parties of wide-eyed tourists, something which aroused the ire of the antiquarian and scholar Johann Joachim Winckelmann, who published a scathing open letter denouncing the practice in 1762.[22] Hence it was only as a result of an express order given by the Secretary of State Bernardo Tanucci in 1764 that more systematic and permanent excavation at the site allowed for the discovery of major public monuments such as the temple of Isis, the gladiators' barracks and two municipal theatres. Even then, the newly arrived British ambassador to the kingdom of the Two Sicilies Sir William Hamilton found only ten to twelve workers aimlessly and randomly digging over the light scoriae covering ancient Pompeii. He complained in a letter to his friend Reverend Doctor William Robertson that 'it would grieve you to see the dilatory and slovenly manner' in which the ruins were being uncovered. Nor were visitors to the excavations permitted to make sketches or drawings of the discoveries, much to the frustration of artists such as Charles Nicolas Cochin, touring the ruins in 1754. Writers such as Goethe, coming to the site as late as 1787, complained that the ancient monuments and houses were still being 'casually ransacked [by their diggers] as if by brigands'.[23]

21. There is a vast bibliography on the discoveries of Herculaneum and Pompeii. For some general works, see Michael Grant, *Cities of Vesuvius: Pompeii and Herculaneum* (London, 2005); *Antiquity recovered: the legacy of Pompeii and Herculaneum*, ed. Victoria C. Gardner Coates and Jon L. Seydl (Los Angeles, CA, 2007); Luciana Jacobelli, *Pompei: la costruzione di un mito* (Rome, 2008); *Pompeii in the public imagination, from rediscovery to today*, ed. Shelley Hales and Joanna Paul (Oxford, 2011); Andrew Wallace-Hadrill, *Herculaneum: past and future* (London, 2012); and Paul Roberts, *Life and death in Pompeii and Herculaneum* (London, 2013).
22. Johann Joachim Winckelmann, *Sendschreiben von den Herculanischen Entdeckungen* (Dresden, n.n., 1762).
23. Hamilton, Cochin and Goethe are cited in the excellent article by Nancy H. Ramage, 'Goods, graves, and scholars: 18th-century archaeologists in Britain and Italy', *American journal of archaeology* 96 (1992), p.653-61 (653-54).

Access to the sites was also a problem. In fact, travellers had first to acquire passports in Rome before embarking on their journey south to Naples. And once in the city further paperwork was required to visit the excavations, specifically the subterranean theatre in Herculaneum. Official permission was also needed to gain entrance to the disparate collections of artefacts removed from the digs and housed in the Royal Museum at Portici from 1755 onwards. By the mid-1770s, according to Neapolitan archives, around a hundred foreign tourists a year were being granted access to the ongoing excavations and to the museum, often accompanied on their visits by their national envoy or by the museum's fiercely proud curator Camillo Paderni.[24] At Portici visitors were captivated much less by the larger pieces of classical sculpture than by the banal, everyday objects preserved by the volcano's action. In 1777 the future revolutionary minister Jean-Marie Roland de La Platière stood entranced before the museum's array of first-century-CE Roman 'pots and pans, bread, fruit, wheat, pasta, fish, wine, weights, measures and scales, musical instruments, writing tablets, dice, incense boxes, make-up, mirrors, scrapers, hairpins [...] rings, bracelets, combs made of wood or horn'.[25] A year or so earlier the marquis de Sade, on the run from charges in France of having poisoned prostitutes and sodomised his valet, was equally enthralled by the sight of foodstuffs seemingly intact after seventeen centuries under the solidified ash and pumice: 'One sees there grain, dates, wheat, peas, beans, pears, a loaf of bread. On the crust is a mark dividing it up into parts, as is still common practice in several countries.'[26] Behind this fascination, common to most visitors to the museum, there is a sense of mortal flesh identifying with perishable foodstuffs, both admiring and envying their preservation over vast swathes of time. Confounding another preconception too, the volcano emerges from the exhibition as much as an agent of benign conservation as one of dark destruction. What is felt dimly in front of the relics of everyday life in Pompeii may also have fed the positive valence of volcanic activity, as this would later be portrayed by groups as diverse as deist philosophers and proponents of louring industrial forges.

24. See Chantal Grell, *Herculanum et Pompéi dans les récits des voyageurs français du XVIIIe siècle* (Naples, 1982), p.5; and Ramage, 'Goods, graves, and scholars', p.656-58.
25. Jean-Marie Roland de La Platière, *Lettres écrites de Suisse, d'Italie, de Sicile et de Malthe*, 6 vols (Amsterdam, n.n., 1780), vol.4, p.234.
26. Donatien Alphonse François de Sade, *Voyage d'Italie* (Paris, 1995), p.252.

Other late-eighteenth-century accounts of the museum's collection dwell censoriously on quite a different human appetite on display. For the excavations in the area had uncovered various phallic amulets, penis-shaped lamps and jugs, pornographic frescoes, not to mention a marble statue of the god Pan pleasuring a she-goat which was thought to be so obscenely graphic that it was only shown to select guests. One English sculptor, possibly Joseph Nollekens, was so struck by the piece that he made a small-scale terracotta replica of it on his return home.[27] However, the most common sexually explicit items in the Portici collection were 'Priapuses'. Named after the Greco-Roman god of debauchery and fecundity, these were usually small grotesque figurines cast in bronze, glass or earthenware representing disproportionately large erect penises. They frequently shocked and intrigued observers such as the abbé Jérôme Richard who perused the collection in 1762. Patently torn between prudishness and prurience, he noted for the edification of his readers: 'A very great form of debauchery was to use a Priapus, made of glass or terracotta, as a drinking vessel', adding that women looking to fall pregnant were made to drink out of them, 'the liquor issuing from the most prominent feature of the Priapus'[28] – that is, his swollen, erect member. Prevalent neo-classical ideals of Roman austerity and virtue as well as the cultured deism of thinkers such as Voltaire, who often praised the tolerant, moderate nature of Roman polytheism, were somewhat unsettled by these finds. For the more righteous, the phallic cults of Campania were deplored as evidence of Rome's irreversible moral degeneration by the first century CE. (Ironically, in Paris in the 1780s group sex acts modelled in wax and sold as boudoir ornaments were doing a brisk trade along the boulevard du Temple and at the big seasonal fairs.)[29]

Among the more horrible and chastening finds made among the ruins of Pompeii and Herculaneum were the remains of those who had perished in the eruption. In the eighteenth century, it was unclear just how many people had lost their lives in the fiery

27. See Roberts, *Life and death in Pompeii and Herculaneum*, p.168-69. The book was published to accompany an exhibition of the same title at the British Museum in 2013, and includes images of both the original statue from the Villa of the Papyri in Herculaneum and the English terracotta copy.
28. Abbé Jérôme Richard, *Description historique et critique de l'Italie, ou Nouveaux Mémoires sur l'état actuel de son gouvernement, des sciences, des arts, des commerces, de la population, et de l'histoire naturelle*, 6 vols (Paris, Delalain, 1770), vol.4, p.495.
29. See François-Marie Mayeur de Saint-Paul, *Le Chroniqueur désœuvré, ou l'Espion du boulevard du Temple*, 2nd edn (Londres [Paris?], n.n., 1782), p.136.

disaster and its chaotic aftermath. We now know that many escaped during the initial hail of pumice and snowfalls of ash as light tephra accumulated at rates of fifteen centimetres per hour, in Pompeii in particular. The body count from excavations at Herculaneum is around 400 out of an estimated population of between 4000 and 5000; in the bigger city of Pompeii, just over 1150 bodies have been recovered from a population of between 12,000 and 15,000, although large swathes of both cities remain encased in volcanic debris.[30] It was originally thought that almost all of Herculaneum had been evacuated, until 300 bodies were found on the shoreline, caught in the infernal hurricane of the first pyroclastic flow around 1am on the second day of the eruption. Their bodies had been blasted by a hurtling avalanche of incandescent rock fragments and superheated ground-soil, ash and cinder, travelling at 70 miles per hour and at 250-400°C. The thermic shock fractured multiple bones, instantly shrivelling sinews into characteristic 'pugilist' poses while, in other victims, brains boiled and burst skulls open.[31] More melancholy discoveries awaited in the buried buildings of Herculaneum, such as a sick boy's skeleton found in a jeweller's shop, lying on an ornate bed with his chicken lunch uneaten beside him. In Pompeii, the finds were equally shocking: sixty gladiators overtaken by the eruption in their barracks, including two bound helplessly in chains; at a wine shop in the Street of Tombs, thirty-four people had taken refuge in an underground chamber, with provisions of food including a goat, but all were suffocated; a further twelve died in the Via Stabiana, after locking themselves in behind an iron shutter.[32] More individual victims were found, crushed by falling masonry or choked to death on air heavy with fine, searing ash and noxious gases, their valuables scattered about them where they fell. Outside the town in the lavish Villa of Diomedes the discovery of twenty skeletons in a cellar in December 1772 caused a sensation, made all the more poignant by the fact that two of them were just boys locked in an eternal, terrified embrace.[33] Although no exact inventory of the bodies was

30. Roberts, *Life and death in Pompeii and Herculaneum*, p.301.

31. Luigi Capasso, *I Fuggiaschi di Ercolano: paleobiologia delle vittime dell'eruzione vesuviana del 79 d.C.* (Rome, 2001), p.21-58.

32. See Grant, *Cities of Vesuvius*, p.17-22.

33. Ernesto De Carolis and Giovanni Patricelli, *Vesuvio 79 d.C. La distruzione di Pompei ed Ercolano* (Rome, 2003), p.121-23. The bodies famously included that of a young woman whose draped breast was perfectly moulded in fine ash. This evocative cast inspired Théophile Gautier's fantastic tale *Arria Marcella* (1852).

kept, it was clear to those excavating or visiting the sites in the mid- to late eighteenth century that the natural catastrophe which had overwhelmed Pompeii and Herculaneum had been on a cataclysmic scale.

As Vesuvius itself was more or less continuously active from the 1750s onwards, with major eruptions punctuating the latter half of the century in 1751, 1767, 1779 and 1794, the revelations from Pompeii and Herculaneum seemed increasingly pertinent, even portentous, to contemporaries. Moved by a mix of neo-classical inspiration and the nascent cult of 'sensibility', as we have seen, visitors often reached for their Greek and Roman poets the better to project themselves into the lives of those who had impotently faced the volcano's wrath (the personification of Vesuvius is a common feature of such accounts). Yet many of the poets and writers they cited had presumed that Vesuvius was extinct. After all, when the rebel gladiator Spartacus led a revolt against Rome in 73 BCE, he and his men hid out on the wooded flanks of the volcano and avoided capture by scaling down the mountain on twisted vines taken from the many vineyards cultivated on its slopes.[34] Virgil was aware of the devastating power of volcanic eruptions, but they were those of Etna, specifically the tremendous explosion of spring 44 BCE which hurled huge columns of smoke and ash into the upper atmosphere, troubling the climate with violent thunderstorms and bizarre cloud formations. It is claimed that this freak volcanogenic weather provided many of the dark omens witnessed before Julius Caesar's assassination.[35] Hence many of the classical visions of volcanism, recalled by educated eighteenth-century travellers to southern Italy, drew explicitly on Etna, Stromboli, Vulcano or Santorini – not quiescent Vesuvius.

34. This episode is recalled by an early-eighteenth-century climber of Vesuvius, Ellis Veryard, in *An Account of divers choice remarks, as well geographical, as historical, political, mathematical, physical, and moral, taken in a journey through the Low-Countries, France, Italy, and part of Spain, with the Isles of Sicily and Malta* (London, S. Smith and B. Walford, 1701), p.217. Veryard claims that seventy-eight gladiators hid with Spartacus in 'the great Pit' of the volcano and escaped their pursuers by abseiling down its sides.
35. See Richard B. Stothers, 'Volcanic dry fogs, climate cooling, and plague pandemics in Europe and the Middle East', *Climatic change* 42 (1999), p.713-23.

The classical and empirical on Etna

Confronted with Vesuvius' terrible subterranean rumblings, Addison recalls the myth of the Titans bucking and tossing to free themselves from their underground prison where Zeus had confined them forever. In this much, he draws on a long-standing mythological cipher to explain the geological phenomena associated with volcanoes. This classical image, however, does not originally concern Vesuvius but derives from Hesiod's assertion in his *Theogony* that the giant Typhoeus, born of Gaia and Tartarus (effectively Earth and Hell) and bent on avenging the Titans, was thunderstruck by Zeus and buried in flames under either Etna or the volcanic island of Ischia off Naples. It is thus the giant's futile, angry struggles to break out of his telluric tomb which generate local earthquakes and volcanic eruptions. The myth was later perpetuated by Pindar, who specifies Etna as the giant's subterranean jail, then by Ovid, before being modified by Milton who replaced Typhoeus with the Titan Enceladus.[36] The myth was clearly still common currency in the late eighteenth century when Sir William Hamilton cited it in passing in his account of the 1779 eruption of Vesuvius sent to the Royal Society in London.[37] Thus many of the classical tropes repeatedly applied by tourists to Vesuvius originated instead in the Ancients' awe before Etna or from encounters with the active volcanoes of the Aeolian Islands off Sicily's north-eastern coast. Virgil, so frequently evoked when beholding the bucolic or infernal landscapes around Naples, famously installed Vulcan's forge under the Aeolian island of Vulcano in the eighth book his *Aeneid.* It was manned by the same colossal, toiling Cyclops who worked the fire god's smithies under Etna in the fourth book of the poet's *Georgics.*

Of all the volcanoes in Europe, extinct, dormant or active, Etna has the longest cultural pedigree. Along with Thera (modern Santorini), Stromboli and the Aeolian Islands, it is regularly cited by ancient historians and writers. It has thus regularly been animated by myths and anthropomorphised by poets (fitting for the only volcano known to blow smoke rings the size of race tracks). In the *Odyssey* Homer

36. David R. Oldroyd, *Thinking about the Earth: a history of ideas in geology* (London, 1996), p.10, 15.
37. Sir William Hamilton, *Supplement to the 'Campi Phlegraei', being an account of the great eruption of Mount Vesuvius in the month of August 1779* (Naples, n.n., 1779), p.16. Hamilton also cites a description of Etna by the Roman poet Martial (p.12).

allegedly located the cave of the Cyclops Polyphemus on the flanks of Etna. This classical reference came repeatedly to the mind of the French art historian Dominique Vivant Denon when he climbed the volcano in 1778.[38] Interestingly, his attempts to rationalise the Homeric myth served only to puncture it. For instance, when he came across a gigantic ancient chestnut grove, petrified on the inside yet still flowering, his thoughts turned to the savage gigantism of the Cyclops who had populated the island in legend. This in turn caused him to perceive vestiges of this mythical race in the taller than average mountain dwellers, in the high numbers of blind among the old, and in the sore, dry, red eyelids of the younger men. However, he ultimately concludes that their red eyes are more probably the result of the 'volcanised air or corrosive, volatile ash that perpetually grinds at the fibres of that sensitive organ'.[39] Seven miles further up the mountain, when his party reaches the so-called Cave of Goats, Denon disabuses his reader of any image that the cave's name might conjure up regarding Polyphemus' fabled lair. Instead, he explains, this scorified recess in the volcanic rock could barely hold six people, and its roof was too low to allow them to stand up in it.[40] Yet, despite Denon's disappointments, the identification of one-eyed giants with volcanic exploration persisted in the nickname given in the eighteenth century to the most celebrated local guides: Blasio on Etna, commonly known as the 'Cyclops', and the yet more famous 'Cyclops of Vesuvius', the literally one-eyed Bartolomeo Pumo, the cicerone recommended by Sir William Hamilton to all his visitors.[41]

Hamilton climbed Etna in June 1769. His account of the ascent, written in October of that year, and dispatched to the Royal Society in London, opens with an alternative classical epigraph drawn from the poem 'Ætna', attributed by Hamilton to the first-century-BCE

38. Dominique Vivant Denon, *Voyage en Sicile* (Paris, Didot l'Aîné, 1788), p.31-32.
39. Denon, *Voyage en Sicile*, p.17.
40. Denon, *Voyage en Sicile*, p.35-36.
41. On Blasio, see Denon, *Voyage en Sicile*, p.33. Canon Giuseppe Recupero, the local expert on Etna, would recommend Blasio to visitors in Catania and they would then seek out the guide in the village of Nicolosi at the start of their ascent. For Bartolomeo Pumo, see the interesting reflections in Noah Heringman, *Sciences of antiquity: romantic antiquarianism, natural history, and knowledge work* (Oxford, 2013), p.109-13; for a fascinating, more extensive analysis of the role of guides on the volcano into the nineteenth century, see also John Brewer, 'The kingdom of Vesuvius', https://brewersblog.org/page/1/ (last accessed 6 December 2018).

writer Cornelius Severus.[42] While the epigraph is little more than a classical flourish, the poem actually argues against superstitious mythological interpretations of the volcano's activities, dismissing popular associations of eruptions with angry giants or the underworld breaking loose. Instead, it anticipates the eighteenth-century taste for analogies drawn from everyday observation and explains Etna's fiery outpourings as the result of what it calls 'lava-stone' heated to explosion by raging underground winds.[43] Lucretius, in *De rerum natura*, had attempted a similar materialist explanation of Etna's volcanism on the model of a furnace, claiming it to be hollow underneath, its subterranean chambers acting like huge telluric bellows to heat and expel air and rock from its crater. According to the Roman writer, the processes of combustion were intensified by seawater being sucked into the interior of the volcano.[44] Hamilton, however, seems less interested in the empiricism of 'Ætna' than in its local colour, as he cites it later in his report to corroborate local lore that the ruins of a tower on the southern slopes of the mountain were those of a temple used by the Ancients to sacrifice to the gods.[45]

The tower in question was, in fact, a major attraction for travellers who scaled the cindery heights. Its chief claim to fame was its association with another exemplary volcanic death – that of the pre-Socratic Greek philosopher Empedocles. According to a legend, revived by the poet Horace in his *Ars poetica* in the first century BCE, Empedocles stepped coolly into the fiery crater of Etna in order to be at one with the elemental forces of fire, air, earth and water, which he believed composed the cosmos.[46] While what has come down to us of Empedocles' writings reveals a more sophisticated philosophy, combining aspects of atomist and evolutionary thinking, visitors to Etna in the eighteenth century were not asked to dwell on the Greek's thought, but were invariably shown the brick and marble remains of the so-called Philosopher's Tower, the Torre del Filosofo, as proof that the sage had indeed set up a crude observatory there in order to study

42. Sir William Hamilton, *Observations on Mount Vesuvius, Mount Etna, and other volcanos: in a series of letters addressed to the Royal Society* (London, T. Cadell, 1772), p.54.
43. *Ætna*, ed. Robinson Ellis (Oxford, 1901), p.48 (on 'lava-stone'). See also Liba Taub, *Ætna and the moon: explaining nature in ancient Greece and Rome* (Corvallis, OR, 2008), especially p.45-55.
44. Lucretius, *De rerum natura* (London, 1955), p.545.
45. Hamilton, *Observations on Mount Vesuvius*, p.82.
46. Horace, *Satires, epistles and ars poetica* (London, 1970), p.488-89.

the volcano's ceaseless evolutions. The French painter Jean-Pierre Houël, who ranged over Etna in 1776, drafting innumerable sketches of the volcanic landscapes, reached the tower in a swirling gale. Shielded against the fierce wind by his companions, he completed a drawing of its crumbling masonry, sunk in what he called the surrounding volcanic 'sands'. However, he was far from convinced of the building's antiquity, stating that its materials and construction had nothing Greek or Roman about them.[47] Denon, for his part, says it is possible that the tower was Roman in origin and could have been constructed as a belvedere for the emperor Hadrian's ascent of the volcano in 125 CE.[48] By contrast, the German nobleman and future diplomat Johann Hermann von Riedesel, who struggled to the summit of Etna in spring 1767, maintains that what remained of the tower's architecture was Norman, dismissing as 'fabulous' his guide's insistence that the ruin once sheltered the Greek philosopher.[49]

One of the most popular accounts of climbing to the crater of Etna was that of the Scottish traveller Patrick Brydone. His picaresque relation of his ascent of the volcano in May 1770 is littered with classical references and empirical observations as well as some vivid tableaux of awe-inspiring sights from the summit, both into the churning lava pit and across the coastal expanses at sunrise. He too cites Odysseus' terrible encounter with the Cyclops Polyphemus, although Brydone situates the dreadful meeting on the southern shore of Sicily, claiming that it was Virgil who later moved it to Etna.[50] As his party scrambled up past the Philosopher's Tower, Brydone leaves the question open as to whether the ruins are those of Empedocles' observatory or remnants of a Roman temple to Vulcan. On his weary descent, however, the traveller is less kind to the memory of Empedocles, recalling Diogenes Laertius' addendum to the myth of the philosopher's elemental union with the volcano, whereby the mountain spat out Empedocles' brass slippers to deny his oneness with the universe and prove his folly in believing himself

47. Jean-Pierre Houël, *Voyage pittoresque des isles de Sicile, de Malte et de Lipari*, 4 vols (Paris, Imprimerie de Monsieur, 1782-1787), vol.2, p.102-103. The drawing of the tower is reproduced between these pages as an aquatint engraving, plate no.122, entitled 'Reste d'édifice *a.* appellé Tour du philosophe'.

48. Denon, *Voyage en Sicile*, p.40-41.

49. Johann Hermann von Riedesel, *Reise durch Sicilien und Großgriechenland* (Zurich, Orell, Gessner, Füesslin, 1771), p.129-30.

50. Patrick Brydone, *A Tour through Sicily and Malta, in a series of letters to William Beckford*, 2 vols (London, W. Stahan; T. Cadell, 1773), vol.2, p.195.

a god.[51] Brydone's engaging tale, combined with news of a powerful eruption of Etna in 1787, inspired a fantastical volcanic episode in a sequel to Rudolf Erich Raspe's *Baron von Munchausen's narrative of his marvellous travels and campaigns in Russia* (1785). The text entitled *Gulliver revived*, published in 1788, recounts how the foolhardy Munchausen jumped into Etna following Empedocles' example, arriving a bit singed at the bottom of the crater, where he finds himself in Vulcan's 'workshop' and meets the god of fire and his Cyclops in the charming company of the goddess Venus. They proceed to give the baron a comical explanation of volcanism before Vulcan becomes jealous of the baron's dalliance with Venus and sends him packing through the centre of the Earth and out into the South Seas.[52] Less fantastical and more whimsical, Xavier de Maistre's anti-travelogue *Voyage autour de ma chambre* (1795) also cites the legend of 'Empedocles' tomb' in the fiery depths of Etna among the destinations of travel writers who recount in tedious detail their ascents, appetites and near-accidents for the titillation of their armchair audience.[53] These humorous references seem to suggest that visitors to the volcanoes of southern Italy were still laden with the cultural baggage of classical culture, but were increasingly inclined to query it, to contrast it with what they experienced in the raw on the barren slopes, even to gently mock its ingenuous or exaggerated representations of volcanic activity.

Curiosity and *katabasis*

The allusions to classical visions of volcanoes indicate the presence of a strong humanist, anti-clerical strain of thinking among those visiting Vesuvius, Etna and their environs in the eighteenth century. This might be seen as an implicit extension of Renaissance culture, discreetly overlooking the Church's traditional association of the volcano with a Christian hell. While occasional references to Vesuvius as a vision of damnation set in an earthly paradise are still to be found in Vigée-Lebrun and Goethe towards the end of

51. Brydone, *A Tour through Sicily*, vol.1, p.183-84, 199-200. For the original tale of the brass slippers, see Diogenes Laertius, *Lives of eminent philosophers*, 2 vols (London, 1925), vol.2, p.382-85.
52. *Gulliver revived: containing singular travels, campaigns, voyages and adventures in Russia, the Caspian Sea, Iceland, Turkey, Egypt, Gibraltar, up the Mediterranean, on the Atlantic Ocean, and through the centre of Mount Etna into the South Sea* [...] *by Baron Munchausen* (Dublin, P. Byrne, 1788), p.121-27.
53. Xavier de Maistre, *Voyage autour de ma chambre* (Paris, 2003), p.74.

the century, Elisabeth Chevallier suggests that 1760 constitutes an effective tipping point towards a neo-classical secularisation of the volcano as a site of philosophical contemplation.[54] More interestingly still perhaps, these late-eighteenth-century accounts of encounters with the active volcanoes of southern Italy point to a clear cultural rehabilitation of the notion of curiosity. At the turn of the century the few tourists to Vesuvius, such as the Englishman William Bromley, might ride without a guide to the volcano, leave their horse at the first lava field, climb to the top of the first foothills, then return to Naples, having 'satisfied [their] Curiosity'.[55] Even those brave enough to make the climb to the summit, such as the Devonshire doctor Ellis Veryard who reached the crater in 1701, were terrified on the ascent, bearing in mind the fate of Pliny the Elder who 'bought his Curiosity at the price of his own life'.[56] Yet the generations of grand tourists and amateur scientists seeking out the volcano after 1760 incarnated a keen inquisitiveness which no longer bore any of the stigma formerly attached to this impulsion. The grand European tradition of seeing curiosity as a vice, especially a feminine one (think Eve, Pandora or the rash heroine of Charles Perrault's 'Barbe bleue'), was increasingly challenged in the eighteenth century. Mme de Scudéry's dictum that 'History is but full of curious people whom Curiosity has undone' no longer dissuaded those straining up Vesuvius, driven instead by the pioneering spirit of their age.[57] Brydone on Etna might warn that its searing crater was treacherous and that those venturing into it have been known to sink into its glowing cinders 'paying for their temerity with their lives'; he nonetheless issued this warning as he leaned eagerly over the crater's rim.[58]

In fact, the classically inspired practice of descending into the crater when the volcano was dormant drew on this new positive sense of curiosity (pushing back the frontiers of knowledge), underwritten

54. Elisabeth Chevallier, 'La finalité du Vésuve: force du mal ou instrument de la providence selon les étrangers venus à Naples au XVIIIe siècle', in *Colloque histoire et historiographie: Clio*, ed. Raymond Chevallier (Paris, 1980), p.349-69.
55. William Bromley, *Several years travels through Portugal, Spain, Italy, Germany, Prussia, Sweden, Denmark and the United Provinces* (London, n.n., 1702), p.129. Bromley did bemusedly report seeing 'a young German Gentleman crawl upon all four to the top' of a higher peak, if not to the crater.
56. Veryard, *An Account of divers choice remarks*, p.217.
57. See Neil Kenny, *The Uses of curiosity in early modern France and Germany* (Oxford, 2004), p.328.
58. Brydone, *A Tour through Sicily*, vol.1, p.196.

by a worldly anti-clericalism (abseiling blithely into hell). Known to the Ancients as *katabasis*, the journey into worlds beneath the surface of the Earth was a staple of Greek and Roman literature. In book 11 of the *Odyssey* Homer's eponymous hero passes into the underground realm to consult the shade of Tiresias. More famously still, Orpheus seeks out his dead lover Eurydice in the subterranean kingdom of Hades, while Aeneas too, guided by the Cumaean Sibyl, makes the downward journey to the underworld to speak to the spirit of his father. These descents were to be brilliantly reinvented in nineteenth-century literature by the likes of Lewis Carroll and Jules Verne,[59] but, before then, they motivated real-life explorations of underground spaces in the eighteenth century. Cave systems in the Harz mountains in Germany were visited by Gottfried Leibniz and Goethe, among others; and caverns in Derbyshire and the Lake District were already established tourist attractions by the late eighteenth century.[60] Underground cavities were also a feature of volcanic exploration. For instance, innumerable tourists to the Neapolitan area were shown the Grotta del Cane near Puozzoli in the Campi Flegrei.[61] The cave was so called because dogs were routinely thrust into its volcanic vapours and shown to asphyxiate there, before being pulled out unconscious and revived by being thrown into a nearby lake. A more humane example would be that of Sir William Hamilton who, roving over Etna in 1769, had himself and his companions lowered on a rope into 'several subterranean caverns' at the foot of Monti Rossi. They were, however, quickly forced to retreat due to the excessive cold and the fierce underground winds that kept extinguishing their torches.[62] In the same text Hamilton also recalls the descent made by Giulio Cesare Braccini into the crater of Vesuvius a few years before its momentous 1631 eruption. Braccini describes wooded terraces teeming with game leading down vast distances to three pools of

59. See Kiera Vaclavik, *Uncharted depths: descent narratives in English and French children's literature* (Oxford, 2010).

60. See Hans-Peter Ecker, 'Vergewisserung subjektiver Identität, Spurensuche im Kollektiven Gedächtnis, Horrortrip in die Wildnis: über kulturelle Programme literarischer Harzreisen', *Zeitschrift für Germanistik* 7:2 (1997), p.262-77; T. A. Hose, 'Towards a history of geotourism: definitions, antecedents and the future', in *The History of geoconservation*, ed. C. V. Burek and C. D. Prosser (London, 2008), p.37-60.

61. See, as one example among many tourist accounts, Charles Pinot Duclos, *Voyage en Italie, ou Considérations sur l'Italie* (Paris, Buisson, 1791), p.389.

62. Hamilton, *Observations on Mount Vesuvius*, p.66-67.

brackish water which he tasted one after the other.[63] Later explorers of Vesuvius were equally keen to take advantage of periods of quiescence to venture into the crater. In May and July 1752, between the lively eruptions of 1751 and 1754-1755, the scholar Giovanni Maria Della Torre went over the volcano's rim and approached what he calls the 'abyss where matter foments, keeping up a bright, continual fire'. There he took measurements of features within the crater and made descriptions of crystals and rock forms, often in the eclectic company of curious guests such as the physicist Jean-Antoine Nollet or the art collector Pierre Louis Randon de Boisset.[64] Similar intellectual curiosity seems to have inspired the worldly monk, intimate of Voltaire and commentator on Newton, père François Jacquier, to be lowered some 200 feet into the crater in the company of a draughtsman in order to sketch the volcano's interior.[65] Nonetheless, from 1760 to the end of the century an intensification of Vesuvius' activity more or less put an end to these secular *katabases* or descents into the underworld. Hence in 1777 Jacques-Joseph de La Roque, having scrambled to the edge of the volcano's 'funnel', as he called it, was forced to forego the descent into the crater by the heat and debris spewing intermittently from it. 'We tried twice', he writes, 'to make this curious descent, but prudence made us think again.'[66]

Returning to where we started this chapter – to the explorations of Vesuvius made by Addison and Watkins at opposite ends of the eighteenth century – we find that both English travellers expressed a similar nagging urge to go down into the steaming crater. But they ultimately held back from making the hot, suffocating descent, possibly restrained by classical example. For the recurrent allusions to Pliny the Elder and Empedocles are nothing if not ancient cautionary tales of two excessive curiosities, two extreme philosophical deaths by volcano. Thus, while Addison thought the crust in the centre of the crater firm enough to walk on, he contented himself with throwing

63. Hamilton, *Observations on Mount Vesuvius*, p.109-10. See Giulio Cesare Braccini, *Dell'incendio fattosi nel Vesuvio a XVI. di Dicembre M.DC.XXXI. e delle sue cause, ed effetti* (Naples, Secondino Roncagliolo, 1632), p.24-27.
64. Giovanni Maria Della Torre, *Storia e fenomeni del Vesuvio* (Naples, Giuseppe Raimondi, 1755), p.17-19.
65. As stated in Sir James Edward Smith, *A Sketch of a tour on the continent, in the years 1786 and 1787*, 3 vols (London, n.n., 1793), vol.2, p.61.
66. Jacques-Joseph de La Roque, *Voyage d'un amateur des arts en Flandre, dans les Pays-Bas, en Hollande, en France, en Savoie, en Italie, en Suisse, fait dans les années 1775-76-77-78*, 4 vols (Amsterdam, n.n., 1783), vol.3, p.48.

rocks onto it.[67] Watkins fought through gagging clouds of acrid vapour just to peer into the enormous vent, brilliant with patches of chlorine and sulphur. He made no attempt to climb down into this simmering pit. However, when his guides shouted, as was their practice, to let him know that the sulphurous steam was clearing so that he could look inside the volcano, their cries were redoubled by a 'loud English huzza' coming somewhat alarmingly from 'a true John Bull' figure triumphant on the rim of the crater. This improbable character was the master of a British merchant ship docked at Naples who had climbed to the summit to glory in 'the most wonderful sight he had ever seen'. Perhaps less inhibited by classical example, the merchantman was, as Watkins put it, 'insensible of the danger' and would have readily descended into the crater 'had I not prevented him'.[68] This anecdote captures neatly the moment when classical restraint meets a more wanton empirical curiosity on the summit of Vesuvius in 1788. The sailor's exertions to reach this 'most wonderful sight' suggest that other discourses on contemporary volcanism, such as an aesthetics of the sublime, also inform his eager curiosity and were not necessarily the preserve of the educated tourist.[69] However, as we shall see, this episode was far from constituting the most reckless and peculiar adventure on the volcano in the course of the century.

Gothic picnics on the volcano: Winckelmann and Sade

On 19 October 1767 Vesuvius erupted violently and spectacularly. Naples was thrown into uproar, with its superstitious poor, the *lazzaroni*, massing in the streets and exhorting the city's patron saint, San Gennaro, to deliver them from the volcano's wrath. The king of Naples, Ferdinand IV, was out hunting near his country palace at Portici, situated on the lower slopes of the erupting volcano, and was forced to flee back to the city. However, at the very same time a small party was beginning its journey in the opposite direction, towards the

67. Addison, *Remarks*, p.240.
68. Watkins, *Travels*, vol.2, p.431-32.
69. Watkins' merchant sailor thoughtlessly seeking to satisfy his untutored curiosity also confirms Nigel Leask's perceptive distinction between 'positive and negative, rational and vulgar, valences of "curiosity"' that emerge in the last decades of the eighteenth century, pointing to the ultimate triumph of a 'commodification and popularization' of curiosity that was to take hold in the early nineteenth century. See Nigel Leask, *Curiosity and the aesthetics of travel writing, 1770-1840* (Oxford, 2002), p.4-5, 23-53 (27-28).

colossal fountains of lava and mushrooming clouds of ash and smoke. This was not a scientific expedition – far from it. The small group comprised the papal antiquary Johann Joachim Winckelmann, the explorer of Etna Johann Hermann von Riedesel and the adventurer Pierre-François Hugues, self-titled baron d'Hancarville, plus their terrified cicerone. In two letters written in late 1767 to friends in Germany, Winckelmann recounts their infernal ascent, relating how he and his companions had frequently to beat their quaking guide to focus his mind on picking a path through the surging lava streams and hot cinders. Finally, with their clothes grimed with ash and their shoes burned and torn, they reached the rim of the crater. There, Winckelmann writes, 'I was the first to undress in order to dry my shirt; my two friends followed suit.' Stripped naked, 'like Cyclops', as the antiquary puts it, the three dilettanti indulged in a truly gothic picnic, roasting dove on the searing lava and washing it down with two or three bottles of Rosoli, a strong and scented local liqueur.[70] For at least one of the party, Riedesel, drinking fortified wine on the flank of a volcano was nothing new. Five months earlier, he had made a gruelling ascent of Etna and had seen his companion on that trip, Emanuel Ferra, abandon the climb several miles from the summit, benumbed and defeated by the icy north winds. Riedesel had only been able to continue by having recourse to 'a bottle of Catanian wine, which is very strong'.[71] Etna is approximately three times the height of Vesuvius, so the debilitating effects of altitude and the persisting snowfields encountered by Riesedel's party meant that their heavy tippling here was the result less of indulgence than of necessity. Patrick Brydone's group of climbers, scrambling up the same volcano exactly three years later and accompanied by the same guide, were similarly fuelled to the summit by great draughts of brandy and at frequent halts 'made a fresh application to [the] liqueur bottle'.[72]

Winckelmann's picnic on erupting Vesuvius might thus appear all the more outlandish and conceited. Yet its trio of dissolute thrill seekers prefigures an even more sensational excursion to the summit of Vesuvius made by a party of libertines in the marquis de Sade's novel *Histoire de Juliette*, published between 1798 and 1801. In Sade's

70. Johann Joachim Winckelmann, *Briefe*, 4 vols (Berlin, 1952-1957), vol.3, p.317. Cited in David Constantine, 'Winckelmann and Sir William Hamilton', *Oxford German studies* 22 (1993), p.55-83 (80).
71. Riedesel, *Reise durch Sicilien*, p.128-29.
72. Brydone, *A Tour through Sicily*, vol.1, p.184.

pornographic and picaresque narrative, the heroine Juliette and her lesbian lover Clairwil invite their fellow debauchee Olympe de Borghèse on a trip to the crater of the volcano. Unbeknownst to Olympe, her libertine companions have become weary of her and plan a spectacularly cruel end to their relationship. On the evening before their outing to the volcano Olympe – whose ironically celestial and mountainous name has not been lost on the reader – dines with her would-be killers. They serve her, of course, a roast. The next day, the three debauchees climb Vesuvius and reach the crater's edge. There Juliette and Clairwil dismiss their guides and immediately set upon Olympe, torturing her sexually before throwing her alive into the volcano's gaping maw. True to the Sadean libertine creed, inevitably aroused by the act of gratuitous evil they have just committed, the lesbian lovers then masturbate together on the crater's rim, their combined orgasm constituting a release of desire that both imitates and parodies a volcanic eruption. Far from offending the natural order, the volcano appears to approve of the human sacrifice made to it, spraying the sated libertines with a harmless shower of small rocks and a belch of smoke. It is as though Vesuvius is happy with their offering, Olympe proving an agreeable morsel swiftly digested in the 'bowels of the earth'. Once again, the libertines choose to mimic the volcano and dig out a picnic from a 'basket of refreshments' that they have had the foresight to bring with them. 'Let nothing disturb us now, let us lunch', Juliette exclaims breezily.[73]

There are obvious parallels between these real-life and fictional episodes: the same-sex trios, the picnics on the top of the volcano, the forced undressing – because of the heat of the lava or because of the heat of blood-lust. The lesbianism in Sade's novel also recalls, albeit obliquely, Winckelmann's own notorious homosexual predilections, something documented by contemporaries such as Casanova and Goethe.[74] Yet, while these correlations might seem to suggest that Sade was aware of Winckelmann's gothic picnic on the volcano, there is only limited circumstantial evidence to support this theory. In effect, the most plausible link is the old libertine writer Ange Goudar, who had lived in Naples in 1767 and later met Sade in Florence in 1775. (Goudar's young English wife Sara is often cited as a model for

73. Donatien Alphonse François de Sade, *Histoire de Juliette, ou les Prospérités du vice*, 2 vols (Paris, 1987), vol.2, p.410-14.

74. See Jacques Casanova de Seingalt, *Histoire de ma vie*, 12 vols (Wiesbaden, 1960-1962), vol.7, p.197-98.

Clairwil.)[75] Yet no written proof has come to light of Goudar relating this story to the marquis. More unlikely, but more tantalising, is the possibility that Sade met d'Hancarville himself in Florence in the same period and so heard this tale from an eyewitness and a participant. This is all the more titillating a supposition, as d'Hancarville was expelled from the kingdom of Naples in 1770 precisely for trafficking in obscene texts, or, as James Moore puts it, for publishing 'explicit pornography depicting men enjoying the intimate company of goats'.[76] Of course, Winckelmann's own gruesome end in June 1768, less than a year after his shared repast on the volcano, is a further anticipation of Sade's fiction: the antiquary was messily stabbed and garrotted in a Trieste hotel room by a male prostitute and apprentice cook, glorying in the name of Arcangeli.[77] But what is perhaps strangest and most telling of all in Sade and Winckelmann's complementary accounts of picnicking on the volcano is that, for all their perverse sensationalism, both represent a well-established contemporary fashion. Picnicking on the fiery slopes of Vesuvius (and to a lesser extent on the more distant, loftier Etna) had, by the 1770s, become a staple part of the Grand Tour, as is amply attested by innumerable travelogues of the time.

Tourist picnics on Vesuvius

In reading tourists' relations of their excursions to the volcano, it is perfectly possible to reconstruct a typical late-eighteenth-century trip to Vesuvius. One would set out early, between 6am and 9am. A light carriage and driver would have been hired in Naples which would cover the six or so miles to Portici. From there, one would usually take the road a further two miles to the village of Resina, the most popular of the three routes practised to the volcano but also, according to Joseph Jérôme Lalande, 'the most difficult'.[78] At Resina, if not before, one would hire mountain guides, although, as the abbé Jérôme Richard warns, one would have to choose carefully from the

75. Maurice Lever, 'Introduction', in Sade, *Voyage d'Italie*, p.5-38.

76. James Moore, 'History as theoretical reconstruction? Baron d'Hancarville and the exploration of ancient mythology in the eighteenth century', in *Reinventing history: the Enlightenment origins of ancient history*, ed. James Moore, Ian Macgregor Morris and Andrew J. Bayliss (London, 2008), p.137-67 (141).

77. See Kevin Parker, 'Winckelmann, historical difference, and the problem of the boy', *Eighteenth-century studies* 25:4 (1992), p.523-44.

78. Joseph Jérôme Lalande, *Voyage d'un François en Italie, fait dans les années 1765 et 1766*, 8 vols (Paris, Desaint, 1769), vol.7, p.167.

crowd of clamouring, dirty wretches who accosted tourists to perform this role, since the local peasantry were 'wicked and self-interested' creatures that it was perfectly acceptable to beat back with a stick.[79] One would also buy torches, the best and cheapest of which were to be found in Naples, writes Jacques-Joseph de La Roque.[80] The torches were needed because tourist parties stayed on the volcano until after nightfall in order to enjoy the incredible nocturnal pyrotechnics of spurting lava and volcanic bombs.

On average it would take four hours to climb to the crater, and a little under three hours to come back down to Resina. The first part of the ascent was undertaken on mule to the edge of the lava flows; from there it was on foot to the summit. Yet if the tourist was exhausted from walking, as Henry Swinburne puts it, 'up to the ancles [*sic*] at every step in purple, lukewarm ashes' or with clambering over jagged old lava beds, then help was at hand.[81] The guides on Vesuvius had developed the technique of attaching weary or unfit tourists to themselves by means of a leather strap tied round their waist; and in this fashion, they would haul the tourist to the summit. This practice is confirmed by Lalande, La Roque and Swinburne, among others, as well as providing a fascinating note of realism to Juliette's fictional ascent of the volcano in Sade's novel.[82]

Nearing the almost always active crater, the tourists' descriptions contain a number of constants: the deafening 'dreadful noise', the bellowing of the volcano likened, as we have seen, to thunder, heavy industry or the reports of the largest artillery. One would rarely fail to mention the unbearable stench of sulphur or remark on one's melted, burned or mangled footwear.[83] In order to arrogate to oneself a degree of the sublime of the object one beheld, it would not be remiss either to pit oneself against the volcano, emphasising the many dangers braved, manifest in the unpredictable showers of molten rock, hissing, rolling or flaming about one's ears or right at one's feet. Charles Dupaty thus excitedly remarks on the 'thousands of stones which whistle, fall,

79. Richard, *Description historique et critique*, vol.4, p.450.
80. La Roque, *Voyage d'un amateur des arts*, vol.3, p.46.
81. Swinburne, *Travels in the Two Sicilies*, vol.1, p.81.
82. Lalande, *Voyage d'un François*, vol.7, p.172; La Roque, *Voyage d'un amateur des arts*, vol.3, p.47; Swinburne, *Travels in the Two Sicilies*, vol.1, p.82; Sade, *Histoire de Juliette*, vol.2, p.411.
83. A fine example of a travel narrative containing these features is Pierre-Jacques-Onesyme Bergeret de Grancourt, *Bergeret et Fragonard: journal inédit d'un voyage en Italie 1773-1774* (Paris, 1895), p.301-303.

rebound and roll – here's one that tumbles down at a hundred paces from me'.[84] While serious injuries were extremely rare on Vesuvius, the bishop of Derry received some scorching volcanic shrapnel in the arm on an early climb with Sir William Hamilton in 1766, which burned him badly, threw him into a fever and had him bedridden for the subsequent five weeks.[85]

As Chantal Grell has clearly shown, tourists to Vesuvius towards the end of the eighteenth century were overwhelmingly English or French, often noble, but increasingly drawn from wealthy professional classes too.[86] As a result, they would not deprive themselves of the pleasures of the table during their explorations of the volcano and its environs. And, as we have noted, the most common form of repast on the slopes of Vesuvius was a kind of picnic. Before examining further what types of food and drink the tourists consumed on the lava-strewn mountainside, it is worth clarifying what is meant by the term 'picnic' in eighteenth-century Europe. In both its French and English forms, the etymology of the term is still hotly disputed, but we do know that the French noun 'pique-nique' was officially recognised by the third edition of the *Dictionnaire de l'Académie française* of 1740, where it is defined as a meal in which 'each pays his part', that is, to which everyone contributes his or her share. This definition – without reference to either a specific locale or a specific menu – is reproduced in the subsequent 1762 and 1798 editions of the same *Dictionnaire*. The word seems to have been used in its English form for the first time in 1748, but does not appear in Johnson's *Dictionary*; it enters the German language as 'picknick' in 1753.[87] Whatever its exact origins, Jean-François Féraud's *Dictionnaire critique de la langue française* assures us that by the 1780s the term 'pique-nique' is 'very commonplace'.[88] Hence it is important to distinguish this contemporary sense of the term 'picnic' from the images commonly evoked of lazy Sunday bourgeois spreads in Parisian parks, inspired particularly by the Impressionist masterpieces by Monet, Manet

84. Dupaty, *Lettres sur l'Italie*, vol.2, p.110.
85. See Alwyn Scarth, *Vesuvius: a biography* (Princeton, NJ, 2009), p.241.
86. Grell, *Herculanum et Pompéi*, p.5-10.
87. See Julia Csergo, 'The picnic in nineteenth-century France: a social event involving food – both a necessity and a form of entertainment', in *Eating out in Europe: picnics, gourmet dining and snacks since the late eighteenth century*, ed. Marc Jacobs and Peter Scholliers (Oxford, 2003), p.139-59 (141-43).
88. Jean-François Féraud, *Dictionnaire critique de la langue française*, 3 vols (Marseille, Mossy, 1787-1788), vol.3, p.166.

and Tissot, all entitled 'Déjeuner sur l'herbe', of the late nineteenth century. A century earlier picnicking did not exclusively signify an outdoor meal; nor was it predominantly associated with one class or section of society. It was characterised instead by a pooling or sharing of dishes and expenses.

In practice, on the lower slopes of Vesuvius, it also meant bringing one's own meal, a habit confirmed by Bergeret de Grancourt in 1774 who writes: 'We had carried our dinner in our carriages to Portici and from there on mules and donkeys to the foot of Vesuvius.'[89] This appears still to be the practice in 1797 when Mariana Starke notes: 'We hired a carriage to go to Resina, [...] took cold meat, bread and a bottle of wine, and set out at daybreak.'[90] In fact, the staple ingredients of these volcanic picnics were habitually cold meats, bread and wine. As for the preferred site for picnicking on Vesuvius, this was a peculiar building called Il Salvatore or the Hermitage, situated about two miles below the crater between Resina and Mount Somma. It appears as a landmark in certain paintings of Vesuvius erupting, such as those by Volaire and Wright of Derby, and the English artist Thomas Jones made a number of sketches of the buildings and its surroundings in 1781.[91] Travellers usually stopped off there on the ascent, but Starke suggests that tourists descending from the summit also broke their journey at this point.[92] According to Lalande, the Hermitage was inhabited by an eccentric Frenchman from Amiens called Claude Velene, who had allegedly deserted the army and sought refuge in this volcanic retreat in the early 1750s.[93] The abbé Richard had met the hermit four years before Lalande in 1762 and – never the most generous of observers – called him 'a compulsive liar who had tried to pass himself off as a miracle worker'.[94] This rather harsh judgement seems, nonetheless, to be borne out by the English tourist Hester Piozzi's visit to this 'good, poor, old man' in the late 1780s when the hermit assured her that he had previously been a hairdresser in London and had

89. Bergeret de Grancourt, *Bergeret et Fragonard*, p.301.
90. Mariana Starke, *Letters from Italy, between the years 1792 and 1798, containing a view on the revolutions in that country*, 2 vols (London, 1800), vol.2, p.130.
91. See Benedict Nicolson, *Joseph Wright of Derby: painter of light*, 2 vols (London, 1968), vol.1, p.76. It is the location today of the Osservatorio Vesuviano, a volcanological research centre built on the ruins of the Hermitage in 1841.
92. Starke, *Letters from Italy*, vol.2, p.131.
93. Lalande, *Voyage d'un François*, vol.7, p.168.
94. Richard, *Description historique et critique*, vol.4, p.453.

undoubtedly curled the hair of some of her party.[95] According to the account of a visit made to the hermitage as late as November 1804 by the German writer and traveller August von Kotzebue, it appears that this affable recluse was still then in residence. Having survived wars and revolutions, his Spartan accommodation had been improved only by the acquisition of a sofa and embellished by charcoal portraits of himself and Napoleon's radical brother Lucien Bonaparte who had recently visited the volcano. Kotzebue also leafed through a visitors' book held by the hermit, mocking the crassness of many of the reflections in it, although owning that many of the most foolish and sentimental comments were written by his German compatriots. One thing, however, remained the same: Kotzebue and his party interrupted their ascent there to lunch on 'cheese, bread, and some very fine *Lacryma Christi* wine that he [the hermit] cultivates himself'.[96]

According to the tourists' accounts, the hermit sometimes provided the food for the picnics; sometimes his guests brought him meals. An anonymous report on the 1767 eruption of Vesuvius sent to the American Philosophical Society claims that tourists who call in at the mountain retreat 'are provided with refreshments by the hermit'.[97] Conversely, the French portraitist Elisabeth Vigée-Lebrun, having fled to Naples from revolutionary France in 1790, states that her visit to Vesuvius involved having 'our dinner brought to the Hermitage where we invited the hermit to share it with us'.[98] Lalande explains that, in fact, both practices were common: 'The hermit plays host to strangers; wine, fruit and some refreshments suited to his state are found there; but those who wish to be well served take care to have provisions brought there.'[99] Whether the tourists chose to dine with the hermit or in the open air with their guides, as Goethe did in 1787, these meals on the volcano offer the image of a very singular sort of

95. Hester Lynch Thrale Piozzi, *Observations and reflections made in the course of a journey through France, Italy and Germany* (Dublin, H. Chamberlaine *et al.*, 1789), p.351-52.
96. August von Kotzebue, *Erinnerungen von einer Reise aus Liefland nach Rom und Neapel*, 3 vols (Berlin, 1805), p.297-300.
97. 'An account of the eruption of Mount Vesuvius in 1767, in a letter from an English gentleman residing at Naples, to John Morgan, M. D. Prof. Med. Coll. Philad.', *Transactions of the American Philosophical Society* 1 (January 1769-January 1771), p.281-85.
98. Louise-Elisabeth Vigée-Lebrun, *Souvenirs*, 3 vols (Paris, 1835-1837), vol.2, p.109.
99. Lalande, *Voyage d'un François*, vol.7, p.168.

eighteenth-century sociability.[100] They represent a commensality – the practice of eating together – which is at once egalitarian, since 'each pays his part', and transgressive, since the wealthy tourists share meals with their social inferiors, that is, with their guides or with the hermit himself. The location of the Hermitage reinforces this implicit equality: perched on the side of the volcano, it is beyond the authority of Church, court or other municipal bodies which regulated dining practices in Naples and its environs. What is more, its proximity to the rumbling, bubbling crater and to Pompeii's ruins provides the hermit's guests with a stark reminder of their shared mortality.

On a deeper, symbolic level, this convivial yet subversive commensality also colours the tourists' subsequent descriptions of the erupting volcano itself. For a start the volcano is frequently personified by means of digestive metaphors: it has, of course, an enormous mouth, or several; it also has 'entrails', as in the recurrent idiom 'the bowels of the earth'. This is consistently found in texts as diverse as Ellis Veryard's atypical account of his ascent of Vesuvius in 1701 and Joseph Wright of Derby's letter to his brother of November 1774, claiming his gaze skimmed the mountain's surface, rather than fathom its volcanic 'bowels'.[101] The volcano is classically defined as *mons ignivomus*, a 'fire-spewing mountain', disgorging lava, pumice, ash and smoke, while this same overflowing lava 'devours' all that it touches.

What is more, the digestive symbolism applied to the volcano is made all the more potent by the profound ambivalence surrounding the nature of fire in the mid- to late eighteenth century. This is clearly articulated in Paul-Henri Thiry d'Holbach's *Système de la nature* of 1770 where 'the matter of fire [matière ignée], which is the principle of life in us, often becomes the principle of our destruction, of cities burned to the ground, of the explosion of a volcano'.[102] Before the revolutionary discoveries of Joseph Priestley and Antoine-Laurent Lavoisier in the 1780s concerning the chemical character of combustion, fire was predominantly understood in terms of the so-called 'phlogiston' theory, whereby ignition and combustion occurred because of the release of each substance's share of phlogiston, a quasi-alchemical essence of fire. The greater a substance's share of phlogiston, the greater

100. See Johann Wolfgang von Goethe, *Italienische Reise* (Wiesbaden, 1959), p.195.

101. Veryard, *An Account of divers choice remarks*, p.216; Wright is cited in Nicolson, *Joseph Wright of Derby*, vol.1, p.10.

102. Paul-Henri Thiry, baron d'Holbach, *Système de la nature* (London, n.n., 1770), p.23.

its flammability.[103] According to this largely 'pre-scientific mindset', fire is strongly imbued with animist qualities.[104] That is, fire suffuses its material host like a living element and, once ignited, it feeds upon its substance. And what applies to a single flame is naturally amplified in the case of red-hot lava and fire-spewing volcanoes. Contemporary scientists, such as the professor of philosophy at Naples University Gaetano d'Amato, were left perplexed by the power of volcanic fire to kindle underground, querying the chemical explanations then in vogue.[105] In his study on Etna and the Aeolian Islands published in 1788, Déodat de Dolomieu thus proposes that lava must possess its own exceptional burning agent, since it retains the power to combust materials that come into contact with it for much longer than do man-made fires.[106] And what confounds the scientific imagination of the period is readily assimilated to atavistic mythical notions among the less educated. In other words, the erupting volcano effects a powerful synthesis of digestive and combustive imagery and in the popular imagination it is easily transfigured into a fiery giant, vomiting all inside it and devouring all around it. If the volcano is animated by the irresistible energy of fire, it is also attributed what Gaston Bachelard calls the 'Pantagruel complex', that is, the insatiable appetite of Rabelais's greedy ogre.[107]

Imagining the volcano as a gigantic fire-breathing glutton has further consequences for its physiological representation. Firstly, its crater is rarely sexualised overtly. It is not feminised as vagina or uterus, that is, as the sexual orifices of Gaia or the reproductive organs of Mother Earth. Instead the crater is configured in terms of ingestion, digestion, regurgitation and excretion. In other words, it becomes the opening of an enormous telluric gut. Thus, when it is not explicitly termed a mouth, it becomes a vast volcanic anus, emptying the earth

103. See, for example, Paul Thagard, *Conceptual revolutions* (Princeton, NJ, 1992), p.39-55; or Eric Guillaume *et al.*, 'Considerations on combustion and fire behaviour of materials: a change of mind during the 18th century', *Journal of fire studies* 34:1 (2016), p.69-84.

104. Gaston Bachelard, *La Psychanalyse du feu* (Paris, 1949), p.109-11.

105. Gaetano d'Amato, *Divisamento critico sulle correnti opinioni intorno a fenomeni del Vesuvio, e degli altri vulcani* (Naples, Stamperia Abbaziana, 1756), p.9, 12-19.

106. Dieudonné Sylvain Guy Tancrède de Gratet de Dolomieu, dit Déodat de Dolomieu, *Mémoire sur les îles ponces, et Catalogue raisonné des produits de l'Etna, pour servir à l'histoire des volcans, suivis de la description de l'éruption de l'Etna, du mois de juillet 1787* (Paris, Cuchet, 1788), p.8-12.

107. Bachelard, *La Psychanalyse du feu*, p.111.

of tons and tons of hot excreta. This also means that attempts to eroticise the volcano in its eruptions lend themselves particularly to representations of those sodomite practices imputed to Winckelmann and extolled in Sade's fiction. Joachim von der Thüsen goes so far as to claim that the volcano is a privileged metaphor for 'that famous center of Sadean sexuality, the anus'.[108] More concretely, Sade himself in his little-studied *Voyage d'Italie* of 1775-1776 relates a curious incident which reinforces this symbolic association of crater and anus. After the terrifying eruption of 1767 the Neapolitan populace erected a makeshift statue of San Gennaro facing the mountain to at once forbid and defy further eruptions. Believing themselves protected by their idol, the vulgar, superstitious crowd, Sade claims, then dropped their trousers and bared their backsides to the volcano, chanting 'Spout, spout, Vesuvius, we defy you! San Gennaro now forbids you!'[109] Yet it was not just the ignorant, brutish Neapolitan mob that stooped to scatological imagery regarding volcanic eruptions. Sir Richard Blackmore had been mocked by Patrick Brydone for diminishing Etna's awesome activity to the notion of 'the mountain [having] a fit of the cholic'.[110] Sir William Hamilton may have employed a more dignified vocabulary in his publications on Vesuvius, for instance describing the evolutions of the powerful eruption of 1779 as the 'fever fits' of the mountain. Yet in his private correspondence he was joyously bawdy, sending Joseph Banks some vulgar local verse on the 1794 eruption entitled 'The mountain of Somma lets down his breeches and shitting addresses his discourse to the Neapolitans'. The eminent volcanologist goes on to assert that, if this last eruption appeared tremendously significant to us, to 'Nature and Mount Vesuvius' it was 'no more than having let out three farts, a sneeze and a Shite'.[111]

From picnic sites to the land of cockaigne

Yet, if picnics on Vesuvius exacerbated the tendency among tourists to characterise the volcano in terms of guzzling, vomiting and defecating, there was another contemporary Neapolitan spectacle that determined

108. Joachim von der Thüsen, 'Juliette at the volcano: reflections on the Sadean journey', *Studies in eighteenth-century culture* 22 (1992), p.87-103 (95).

109. Sade, *Voyage d'Italie*, p.247.

110. Brydone, *A Tour through Sicily*, vol.1, p.249.

111. Hamilton, *Supplement to the 'Campi Phlegraei'*, p.8, 14-15; cited in David Constantine, *Fields of fire: a life of Sir William Hamilton* (London, 2001), p.197.

even more profoundly this identification of the volcanic with the alimentary. This was the 'cuccagna', or 'cockaigne' in English, a term deriving from the medieval peasant's fabled 'land of cockaigne' or land of plenty. An integral part of almost all popular festivities in eighteenth-century Naples, the cockaigne took place in the large square called the Largo del Castello, opposite the royal palace and often in the presence of the royal family and assorted Neapolitan nobility. Again in his *Voyage d'Italie* Sade gives a vivid description of this grotesque public feast:

> The carnival opens with a cockaigne, the most barbaric spectacle that it is possible to imagine in the world. On a large scaffold decorated with rustic trimmings, there is placed a prodigious quantity of foodstuffs arranged in such a way as to compose a part of the decoration themselves. There, inhumanly crucified, are geese, chickens, turkeys that, being pinned up alive with a couple of nails, entertain the populace with their convulsions until it is time for the people to go and pillage all of this.[112]

One of the abbé de Saint-Non's correspondents witnessed a similar cockaigne in 1778 and reported that the construction was covered with 'dried fish, loaves of bread, quarters of beef, mutton, veal and geese that are barbarously strung up, often four days in advance'.[113] Once these gruesome preparations were complete, at midday precisely a cannon was fired from the Castello signalling the start of the ferocious pillage of the cockaigne. According to eyewitness reports, the ravenous *lazzaroni* attacked the amassed sides of meat, live poultry and other foodstuffs with such unbridled voracity that the structure was sometimes stripped bare in less than five minutes. Nothing and no one was spared in the popular feeding frenzy, with the live birds ripped from their fastenings so violently as to 'leave their wings behind', while it was not uncommon for a good number of the feasting mob also to be left maimed or even killed in the greedy stampede.[114]

This might then seem a far cry from the light convivial picnics enjoyed on the neighbouring volcano. But reminders of Vesuvius were

112. Sade, *Voyage d'Italie*, p.177.

113. Jean-Claude Richard de Saint-Non, *Voyage pittoresque, ou Description des royaumes de Naples et de Sicile*, 5 vols (Paris, Clousier, 1781-1786), vol.1, p.250.

114. See Saint-Non, *Voyage pittoresque*, vol.1, p.249; Lady Anna Riggs Miller, *Letters from Italy, describing the manners, customs, antiquities, paintings, etc. of that country in the years MDCCLXX and MDCCLXXI to a friend residing in France*, 3 vols (Dublin, Watson *et al.*, 1776), vol.2, p.178; Sade, *Voyage d'Italie*, p.178.

never far away. For a start, the heaped structure of the Neapolitan cockaigne recalled the looming mountain smoking in the background. Whereas elsewhere in Europe at the time, a cockaigne was usually a tall slim mast with a leg of meat hoisted on top, in Naples it was physically a mountain of food. Lady Anna Riggs Miller describes the cockaigne she attended in February 1771 as a wooden construction, 'the greater part of which is covered over with different articles of provision', forming 'a hill [...] exactly opposite the palace from the center of which bursts forth a fountain, and falls into a bason at its foot'.[115] What else would this overflowing monticule call to mind but an erupting volcano? What is more, the explosion of hungry savagery starts with a cannon shot, recalling the booming artillery used as a rhetorical figure by tourists to describe the thundering noises heard on the slopes of Vesuvius. And like the volcano, the ravenous crowd erupts into action, destroying all in its path. 'The cannon smoke has not dissipated before every last scrap has been carried off', writes Saint-Non's correspondent.[116] Nonetheless, dynamically at least, the cockaigne inverts the movement of a volcanic eruption. Whereas Vesuvius' lava overflows the crater and spills down the sides of the mountain, the *lazzaroni* pour up the sides of the cockaigne, stripping it to its summit. Whereas the volcano disgorges, the people gorge themselves. Yet in both instances, volcano and cockaigne, lava and *lazzaroni* share a spectacular intemperance of digestion and regurgitation, a terrifying bulimic excess.

For many of the visitors and tourists witnessing both of these eruptions – of volcanic might and popular greed – it is possible that a conscious or unconscious link is forged between the two experiences, both presented as spectacles and centred on eating or figures of eating. Certainly, in the eyes of its visitors, eighteenth-century Naples was the epitome of the spectacular. Situated on a stunning bay in the form of a natural amphitheatre, the city was famous for the open-air theatricality of its inhabitants, for its magnificent San Carlo opera house, grand baroque churches and imposing palazzi, so that many tourists would echo Hester Piozzi's sense of visual surfeit when she writes that her 'eagerness to see sights has been repressed at Naples only be finding everything a sight'.[117] The newly unearthed ruins of Herculaneum and Pompeii only confirm this impression of shameless spectacle, boasting a

115. Miller, *Letters from Italy*, vol.2, p.176.
116. Saint-Non, *Voyage pittoresque*, vol.1, p.250.
117. Piozzi, *Observations and reflections*, p.307.

large forum, public theatre and amphitheatre among their sensational finds. Vesuvius too, of course, was emblazoned on the skyline as a further unmissable sight and, especially when it was active, stood as an attraction to be hungrily consumed by tourists and travellers to the region. This implicit understanding of the volcano as at once unique entertainment and sumptuous delight for the senses is captured in Alberto Fortis' *Saggio d'osservazioni sopra l'isola di Cherso ed Osero.* In the early summer of 1771 the Italian geologist was due to explore the eastern Adriatic islands with his English patron the bishop of Derry (the same person who had been injured by red-hot volcanic debris in 1766 in the company of Sir William Hamilton). But at Pula the two travellers were greeted by the exciting news of a 'fresh eruption of Vesuvius' providing them with 'the strongest temptation' to revise their plans. As a result, Fortis writes, 'we crossed from Rovigno [Rovinj] to Ancona, in the hope of enjoying a portion of the feast; but at Rome we were given the unwelcome news that the mountain had fallen quiet.'[118] As we have established, the volcano itself vomits and devours in gargantuan fashion, but, in doing so, it paradoxically also becomes a delicious object of consumption itself, a spectacular 'feast' for those touring southern Italy in the latter part of the eighteenth century.

A further connection between the volcano and contemporary eating practices is even more paradoxical. It concerns the early modern fad for confecting and consuming ices in southern Italy which gave rise to a lucrative trade in mountain snow. The fashion for ices dates from the late seventeenth century and is celebrated in Antonio Latini's *Lo scalco alla moderna: overo l'arte di ben disporre li conviti* (1692-1694). Latini was major-domo to Don Stefano Carillo Salcedo, first minister of the viceroy in Naples; and in this role he provides ample evidence for the popularity of ices at the Spanish court.[119] Both water-based sorbets and milk-based ice creams, in a large variety of flavours, were usually served as an early evening speciality. Such was the consumption of ices in the offices of the neighbouring government in Palermo that by the 1710s it was mockingly known as the 'ice-cream parliament'.[120]

118. Alberto Fortis, *Saggio d'osservazioni sopra l'isola di Cherso ed Osero* (Venice, Gaspare Storti, 1771), p.148.

119. Elizabeth David, *Harvest of the cold months: the social history of ice and ices* (London, 1994), p.141-48; Jeri Quinzio, *Of sugar and snow: a history of ice cream making* (Berkeley, CA, 2009), p.9-15.

120. See Denis Mack Smith, *A History of Sicily: modern Sicily, after 1713*, 2 vols (London, 1968), vol.2, p.266.

The connection with volcanoes was the deep snows that covered their peaks in the winter months and which constituted an important source of revenue and trade for the Spanish government in Naples. Visiting the region in 1701, Joseph Addison relates that the government issued monopolies to 'Snow-Merchants' ('nevaioli') who guaranteed the city a steady supply of snow and ice over the year at a settled price. They oversaw the local poor who dug pits high in the mountains to the east of Naples on volcanic Mount Somma, although the principal source of snow appears to have been Monte Faito above the port of Vico Equense on the southern shore of the Bay of Naples.[121] Snow was rolled into the pits, rammed into solid blocks of ice and covered against the sun. The blocks were cut as and when demand warranted it, when they would be packed in layers of salt and straw, loaded on asses and conveyed to sea ports from where they were shipped to the large coastal cities for sale and consumption. Yet, Addison remarks, even in the early eighteenth century, racketeering was rife in the Neapolitan region, and the snow-merchants were often victims of extortion from local *banditti* (forerunners of today's Camorra) who demanded protection money on threat of destroying the snow reserves 'by the Infusion of some Barrels of Oil'.[122]

Mixed with salt and packed tightly around metal moulds, snow would freeze and preserve the contents or it would be used to cool wine and liquors. Sugared, it formed the basis of a bewildering range of sorbets, ices and ice creams. Vincenzo Corrado's *Il cuoco galante*, published in Naples in 1773 and selling over 4500 copies in various editions from its publication to 1786, offers recipes for cold cream desserts that include the flavours lemon, strawberry, coffee, peach, vanilla, orange, cinnamon, chocolate, pistachio, melon, candied fruit, sage and at least one called 'alla neve'.[123] Ices were also promoted for their salutary attributes. Not only did they slake summer thirsts but they were also thought to alleviate overheated stomachs, allay fevers and reduce the risks of paralysis and scurvy.

121. Addison, *Remarks*, p.242-43; for a brilliant exposition of the snow trade in eighteenth-century Naples, see Melissa Calaresu, 'Making and eating ice cream in Naples: rethinking consumption and sociability in the eighteenth century', *Past and present* 220 (August 2013), p.35-78.

122. Addison, *Remarks*, p.244.

123. See Vincenzo Corrado, *Il cuoco galante* (Naples, Stamperia Raimondiana, 1773), p.115-18; and Calaresu, 'Making and eating ice cream in Naples', p.62-63, which gives details of Corrado's sequel to his bestseller, entitled *Il credenziere di buon gusto* (Naples, n.n., 1778), containing a longer section on ice-cream making.

The Neapolitan author Filippo Baldini wrote an entire treatise in 1775 on the medicinal qualities of different sorbets, claiming that they could relieve everything from melancholy to diarrhea.[124] Summering in Palermo, Patrick Brydone was an advocate for the health benefits of ice and claimed 'it braces the stomach and gives a new tone to the whole fibres'.[125] And, in effect, the important trade in mountain snows at Naples was yet more substantial in Sicily where Etna provided vast quantities for local and international commerce. Riedesel notes in 1767 that the villagers living on the higher slopes of the volcano 'earn their livelihood by transporting the snow to Catania and Riposto, a village on the seashore, where Maltese ships come to load it up for their island'.[126] The snows of Etna were harvested and held in high mountain caves, especially on the northern flanks of the volcano. Yet, as Brydone remarks acidly, if the local inhabitants did the bulk of the physical labour involved in stocking and shipping the snows, the principal benefactor of the Sicilian snow trade was the bishop of Catania who issued the rights to trade in it. One small northern section of the volcano was said to fetch the bishop an annual revenue in excess of 1000 *livres*, a huge sum at the time.[127] According to Henry Swinburne, the Jesuit order in Naples proved equally entrepreneurial in exploiting their exclusive privilege to supply the city with snow.[128]

Notwithstanding the corruption or exploitation associated with harvesting volcanic snows in eighteenth-century Italy, travellers including Addison, John Moore, Brydone and Swinburne insist on the immense popularity of ices across all orders of society, greedily consumed by *lazzaroni* and nobles alike.[129] In fact, at Naples as more generally in Sicily, it was a commonplace to state that 'a Scarcity of Snow would raise a mutiny [...] as much as a Dearth of Corn or Provisions in another Country.'[130] In other words, eating was highly political; and the volcano was key to understanding the central place that food occupied in the Neapolitan or Sicilian social orders of the time. Not only were the snows harvested on its slopes important ingredients of local commerce and consumption but, as we have

124. Filippo Baldini, *De' sorbetti: saggio medico-fisico* (Naples, n.n., 1775), cited in Calaresu, 'Making and eating ice cream in Naples', p.61-62.
125. Brydone, *A Tour through Sicily*, vol.2, p.214.
126. Riedesel, *Reise durch Sicilien*, p.135.
127. Brydone, *A Tour through Sicily*, vol.1, p.139.
128. Swinburne, *Travels in the Two Sicilies*, vol.1, p.105.
129. See Calaresu, 'Making and eating ice cream in Naples', p.50-61, 66-78.
130. Addison, *Remarks*, p.242.

seen, the volcano also stood as a fire-breathing metaphor for the insatiable alimentary energies released by popular festivities such as the cockaigne. Thus, if the erupting volcano and the frenzied cockaigne represent two symbolic extremes of disgorging and gorging respectively, they also stand as the limits between which the food-obsessed social imaginary at Naples is seen to function.

Incommensurability and measure

Eighteenth-century tourists to the Neapolitan region thus found themselves in the realm of a double excess, face to face with two outsized appetites: that of Vesuvius and that of the people. Rather than being exceptional, the immoderate diets of both the volcano and the *lazzaroni* seemed endemic to Campania itself. Visitors to the royal museum of antiquities at Portici were similarly struck by the mounds of perfectly petrified foodstuffs and the racks of kitchen utensils recovered from the ruins of Pompeii and Herculaneum. The gourmand Bergeret de Grancourt was glad to see that the Romans shared his interest in the pleasures of the table: 'One is pleased to see that they fully satisfied the same needs as us.'[131] As we have already seen, the museum also contained reminders of a further excessive appetite among the ancestors of eighteenth-century Neapolitans: their explicit sexual immoderation, manifest in the collection of obscene votaries and amulets carved or sculpted in the shape of enormous phalluses. These priapic charms at once shocked and intrigued prurient tourists such as the abbé Richard or Lalande, and seemed to constitute further proof that the volcano and its people were animated by a common, incommensurable force of nature.[132]

So what was the tourists' reaction when faced with displays of natural and social energy that they were clearly incapable of controlling or comprehending? The general response seems to have been a variant on the approach adopted by the most famous foreigner in Naples at the time, Sir William Hamilton, the British ambassador to the kingdom of the Two Sicilies and a pioneering volcanologist. Knowing himself to be largely powerless to tame the cataclysmic might of the volcano or the irrepressible wildness of the Neapolitan mob, Hamilton chose to observe their respective violence from as close as possible. Then, once the periodic explosion was past, he

131. Bergeret de Grancourt, *Bergeret et Fragonard*, p.321.
132. See Grell, *Herculanum et Pompéi*, p.134-39.

would measure, gauge, evaluate and draft a report on it – to be sent either to the Royal Society or to the British Foreign Ministry. Yet, as Noah Heringman has pointed out specifically in relation to Hamilton's volcanology, his approach here is fundamentally flawed, since measuring volcanic tephra (the size of boulders, the extent of lava flows) gives only a very faint idea of the enormous volcanic energies involved in heaving them out. Heringman argues that, in order to compensate for the *après-coup* character of his measured and measuring reports, Hamilton is increasingly obliged to reduce as far as possible the spatial and temporal interval separating himself from the eruption.[133] The upshot of this is that in the course of his thirty-six years in post in Naples, the dauntless volcanologist embarked on ever riskier expeditions among treacherous lava flows, under hails of hot pumice and molten rock and through choking sulphurous fogs. In alimentary terms, the danger he constantly ran was of being 'eaten' by the volcano, swallowed up in its bubbling crater or engulfed by its spouting lava. And, from his initial forays among the spitting rocks and scorching lava flows in 1766, Hamilton conceives of the dangers he courts in terms of ingestion: 'I was apprehensive of the opening of a fresh mouth, which might have cut off our retreat', he writes in his first report to the Royal Society.[134] Hamilton's writings on Vesuvius are the most immediate and intrepid accounts of volcanological exploration that we have from this first age of geological fieldwork. Yet they remain self-consciously measured and circumspect. This is in part because Hamilton sought scientific objectivity in his approach; but also because, as an enlightened observer, he believed that 'measure', moderation, was the appropriate response to the extreme displays of truculence and violence around him. (The aesthetic corollary of this scientific measure is to be found in Hamilton's collection of simple but beautiful Etruscan and Greek vases, many recovered from the excavations at Pompeii and Herculaneum; they in turn confirm the philosophical measure of the classical Greek mean, a tempering value dear to Hamilton the diplomat.)[135]

A sense of measure could also be understood literally here. In the face of overwhelming volcanic forces, Hamilton reaches for the

133. Noah Heringman, 'The style of natural catastrophes', *The Huntingdon Library quarterly* 66:1-2 (2003), p.97-133 (125-26).

134. Hamilton, *Observations on Mount Vesuvius*, p.27.

135. See *Vases and volcanoes: Sir William Hamilton and his collection*, ed. Ian Jenkins and Kim Sloan (London, 1996).

latest scientific tools with which to gauge, and thereby in some small way delimit, the devastating power of an eruption. In this he is aided by a renewed contemporary interest in metrology, or the art of measurements, as well as by significant technical advances in the manufacture of scientific instruments. In 1779 he explicitly cites the 'most excellent refracting telescopes' that allow him to approach the towering columns of smoke and ash issuing from the erupting volcano when physical access to them is simply unthinkable or downright suicidal. Other volcanologists of the age refer to their instrumentation too, as Lalande, accompanied by Della Torre, scrutinises a 'barometer' at frequent stops in their exploration of Vesuvius or as Brydone incessantly checks the Fahrenheit thermometer, 'made by Adams in London', in his ascent of Etna, and on reaching the summit scans the dawn skies with 'a small glass which I had in my pocket'.[136] More widespread still is the seemingly irresistible urge to throw stones into the brewing crater and time their descent with pocket watches or simply by counting off the seconds. This is an almost universal method of calculating roughly the crater depth, and is wickedly parodied in Sade's *Histoire de Juliette* when the two lesbian murderesses time the fall of the unfortunate Olympe de Borghèse's body into the lava pit. Deliciously, 'for more than six minutes' it thuds, rips and spins off the spurs of rock lining the crater.[137] Ultimately, of course, like the various forms of aesthetic domestication applied to the volcano, the systematic and increasingly accurate measurements taken of the volcano's lava flows and boulder-like ejecta are intended to make the inscrutable workings of volcanism submit to human understanding. They represent necessarily belated and corollary means of apprehending (if not always comprehending) the unfathomable energies unleashed by an eruption.

In broadly sociological terms, the tourists' exultant accounts of climbing Vesuvius possibly fulfil a similar function to the scientists' more specific measurements of lava flows and calculations of the crater's width and depth. Namely, the texts also seek to restore some measure or order to the world after their authors have experienced the volcano or the *lazzaroni*'s dreadful, dizzying, chaotic violence up close. The accounts are attempts by the writers to regain a certain collective control over the natural environment or to reassert self-control in the face of forces beyond their wildest imaginings. The problem is that

136. Hamilton, *Supplement to the 'Campi Phlegraei'*, p.6; Lalande, *Voyage d'un François*, vol.7, p.169-70; Brydone, *A Tour through Sicily*, vol.1, p.105, 185-87.

137. Sade, *Histoire de Juliette*, vol.2, p.412-13.

they only ever do so *après-coup*, after the event, so to speak. This is not the case with the picnics spread on the rim of the fiery crater or on the volcano's lava-streaked slopes. These acts of communal eating constitute a more immediate defiance of the volcano's awful might. That is, they restore a certain proportion and measure to the world by giving an example of moderation – alimentary, dietary this time – in the face of the outsized, insatiable appetites embodied by the Neapolitan people and their neighbouring volcano.

There is something quietly heroic about picnicking on the volcanic precipice. Roland de La Platière's second ascent of Vesuvius involved an example of such modest heroism. Once his party had braved the sulphurous blasts of the crater edge, they settled in the lee of the wind 'each perched as best he can, as the meat is distributed and the flagons emptied'. (In fact, the Frenchman was so struck by this picnicking experience that he returned to the volcano with a friend from Lyon a few days later and, encamped on a 'terrasse' of the boiling-booming crater, they tucked into 'cold meats, bread, wine in quite substantial quantities'.)[138] This confirms what Claude Grignon has stated about picnics in general: that they belong to the order of 'transgressive commensality [...], which, through symbolic compensation and inversion, allow[s] the ordinary order of things to be accepted anew and to resume'.[139] They are, in other words, a more polite and civilised way of baring one's backside to the volcano. But more significantly here, Roland notes that his volcanic picnics are reminiscent of the classical heroes of Homer and Virgil who celebrate any victory over death by first sitting down to table, for 'there is nothing like feasting for forgetting tiredness and danger'.[140] He thus neatly combines classical and sociable eighteenth-century responses to the raw encounter with an active volcano.

Nonetheless, in the course of the century, it would appear that the balance between the conception of volcanism formed on a classical example and one born of empirical exploration tipped in favour of the latter. This is in part because other discourses on volcanoes queried the Ancients' accounts, even as these same discourses recited them. The valorisation of subjective feeling, experienced as exhilarating

138. Roland de La Platière, *Lettres écrites de Suisse*, vol.4, p.32, 40-41.

139. Claude Grignon, 'Commensality and social morphology: an essay in typology', in *Food, drink and identity: cooking, eating and drinking in Europe since the Middle Ages*, ed. Peter Scholliers (Oxford, 2001), p.23-33 (31).

140. Roland de La Platière, *Lettres écrites de Suisse*, vol.4, p.31-32.

terror on the crater rim or delicious horror when confronted with the skeletons unearthed in Pompeii, brought the volcano thrillingly into the present. This modern sensibility also informed a quest for local colour in one's depictions of volcanic terrain, often accentuated by the prevailing aesthetics of the sublime and the picturesque. And, as we shall see in the next chapter, empirical enquiry into the causes of volcanism disdained mythological explanations, as nascent earth sciences developed their own speculative theories and *in situ* protocols of observation on volcanoes, often drawing on scientific advances in cognate disciplines.

2. Two modern 'Plinies' and the empirical turn

After more than three centuries of quiescence, Vesuvius erupted with catastrophic violence in December 1631. Pumice and ash were pumped twenty-one kilometres into the atmosphere and lower stratosphere in a classical Plinian column which was blown as far east as the Balkans and Constantinople. The eruption triggered heavy thunder-storms and localised earthquakes, which generated roiling mudflows, known as lahars, on the volcano's flanks. These mudflows, freighted with steaming ash, rock fragments and huge boulders, swept through almost all human habitation at the foot of the volcano, killing as many as 10,000 people – up to five times the number of victims claimed in ancient Pompeii. A further 40,000 terrified souls fled to Naples to seek refuge, many overcome by ash or debris flows on the way. As the earth shook in Naples, the cone of the volcano collapsed into the vent, forcing out pyroclastic flows which scythed down around 2000 people in Torre del Greco on the coast. Their charred remains were swept out to sea. Yet many corpses would return, as the eruption caused tides to recede suddenly, as though the sea were inhaled by the volcano itself, before the waters surged back in a terrible tsunami five metres high, which smashed repeatedly through fishing villages along the Bay of Naples, taking yet more lives. Two days after the volcano's initial blast, its activity had diminished enough to allow the Spanish Habsburg authorities to send out scouting parties. They discovered victims roasted alive in two metres of smouldering ash and dogs eating burned and scattered human remains. The mountain itself did not escape unscathed from the eruption: its crater was enlarged to almost three times its previous width, and it was shorn of almost 500 metres in height.[1]

1. Scarth, *Vesuvius: a biography*, p.135-69; also Sean Cocco, *Watching Vesuvius: a history of science and culture in early modern Italy* (Chicago, IL, 2013), p.52-78. On the

Etna too was active in the 1630s, spurting into fiery life in 1637, as it did again in 1651. However, these were but preludes to the cataclysmic eruption of March 1669, which spouted huge fountains of red-hot rock and cinders from the mountain's lower slopes just above the village of Nicolosi. The eruption also threw up the double cone of Monti Rossi and spewed out a river of lava up to two miles wide, spilling down to the medieval city of Catania more than twelve miles away. A group of Catanians, led by the Maltese knight and local priest Diego Pappalardo, made brave attempts to break through the banks of solidifying lava and divert the flow away from the city. But ultimately they could not prevent the gigantic, pulsing torrent of molten rock from breaching the city walls.[2] By late April, the lava had overwhelmed the west of Catania and reached the port where 'this matter like ragged rocks, burning in four fathom water, two fathom higher than the sea itself', poured into the bay, 'causing a great and horrible noise, smoak and hissing in the Sea'. So writes the third earl of Winchilsea, British ambassador to Constantinople, then returning to Britain via the Strait of Messina, who was an astonished eyewitness to the eruption. He estimated that the steaming lava advanced up to 600 yards into the sea in a sizzling mile-wide front.[3]

On the influence of Kircher and chemistry

Confronted with this spectacular, violent volcanic activity, seventeenth-century science elaborated a number of original explanations for its occurrence. One of the most influential of these came from a German Jesuit scholar who had witnessed in person the eruption of Etna in March 1637. Athanasius Kircher was a man of his time:

transformation to the volcano's form, see Antonio Nazzaro, 'Implicazioni di una ermeneutica delle fonti vesuviane sull'eruzione del 1631: forma del vulcano e risposta al rischio', in *Napoli e il Gigante*, ed. R. Casapullo and L. Gianfrancesco, p.137-78. The most famous eyewitness account is that of Giulio Cesare Braccini, *Dell'incendio fattosi nel Vesuvio.*

2. See Francesco Ferrara, *Descrizione dell'Etna con la storia delle eruzioni e il catalogo dei prodotti* (Palermo, 1818), p.101-10. Ferrara draws on two important contemporary eyewitness texts: Giovanni Alfonso Borelli, *Historia ed meteorologia incendii Aetnaei anni 1669* (Naples, Egidio Longo, 1670), and Tomaso Tedeschi, *Breve raguaglio degl'incendi di Mongibello avvenuti in quest'anno 1669* (Reggio di Calabria, Domenico Ferro, 1669).

3. Heneage Finch, earl of Winchilsea, *A True and exact relation of the late prodigious earthquake & eruption of Mount Aetna, or Monte-Gibello* (London, T. Newcombe, 1669), p.4-5.

a contemporary of Descartes, Boyle and Kepler, and, like them, a scholar caught on the cusp of two worldviews. A Catholic mystic and a Renaissance polymath, his range of interests was staggering: baroque music, Egyptology, magnetism, symbolic languages, sundials and the plague, to name but a few. He was a writer of speculative prehistories of Noah's Ark and the Tower of Babel, but also the author of a study of the ruins around Rome where from 1634 he studied and worked in the headquarters of the Jesuit order. His mind was a baroque museum in which gryphons, mermaids and angels cohabited with mechanical toys, magnetic experiments and rock samples. Yet it is in the last of these fields, in the earth sciences, that his holistic worldview most successfully accommodated empirical proofs and first-hand observations.[4]

Kircher's key contribution to conceptualising the workings of the Earth was his *Mundus subterraneus* of 1664-1665. This was in large part inspired by his journey through southern Italy and Sicily in 1637 and 1638 as father-confessor and travelling companion to Friedrich, landgrave of Hesse-Darmstadt. There, in addition to witnessing the impressive spectacle of Etna erupting, Kircher clambered over Stromboli and Vulcano. On the return sea crossing in March 1638, he experienced a terrifying earthquake off Calabria and barely made it ashore. Arriving at Naples, he wasted no time in visiting Vesuvius. He made an arduous night-time ascent with a local guide, his 'rustic companion', who led him directly to the crater. Kircher's reaction was typically confounding: he uttered an awestruck Latin prayer, imagined seeing 'Vulcan's kitchen at work', and took extensive measurements and observations of the rumbling, hissing crater. As dawn broke, he had himself lowered into the boiling pit of Vesuvius where he took out his 'Pantometer', or universal measurer, in order to gauge the compass and depth of the newly enlarged opening.[5] While his final measurements are fanciful, they connote important aspects of Kircher's conception of volcanism. Firstly, it is dynamic:

4. On Kircher, see Joscelyn Godwin, *Athanasius Kircher: a Renaissance man and the quest for lost knowledge* (London, 1979); and *Athanasius Kircher: the last man who knew everything*, ed. Paula Findlen (New York, 2004).

5. Athanasius Kircher, *Mundus subterraneus* (Amsterdam, Joannem Janssonium & Elizeum Weyerstraten, 1665), 'Præfatio', ch.3, not paginated. See also Chantal Grell, 'Regarder, observer et interroger le Vésuve: le volcan comme "objet d'étude" au XVIIIe siècle', in *Le Vésuve en éruption*, ed. E. Beck Saiello and D. Bertrand, p.51-81, which includes a French translation of Kircher's reflections on Vesuvius, p.78-81.

the volcano is an active, shifting geophysical feature, as evidenced by the rising and sinking central cone formations witnessed by the Jesuit. Secondly, it is a deep phenomenon: Kircher conceives of the volcano as a roaring telluric chimney-vent or, in an anatomical metaphor, the 'viscera' of the Earth. These two analogies – to architecture and the body – converge in their construction insofar as both are constituted around the channelling of vital flows, be they streams of fire or the circulation of blood and humours.[6] The old Renaissance correlation between microcosm (the human body) and the macrocosm (the Earth) is reactivated geologically by Kircher's experience of the volcanoes of southern Italy.

One of Kircher's reasons for examining the crater of Vesuvius was to see whether it afforded any evidence of secret communication with Stromboli and Etna, as his own theory of the Earth maintained. For he extrapolated from his observations that the Earth had at its core a central fire which fuelled a radiating series of interconnected canals, chambers and conduits; and where these channels breached the surface they formed volcanoes.[7] These subterranean fires were especially volatile and impetuous where they intersected a similar underground network of water courses. In the wake of the tremendous explosion of Etna in 1669, an anonymous English text explicitly drew on Kircher's theories in order to interpret Winchilsea's account of the eruption. It concludes, as the Jesuit suggests, that God as cosmic architect uses volcanoes as part of an elaborate, dynamic system of global homeostasis. Thus it states: 'You see therefore the manner and the way of the Circulation of Nature. You see how Water, Fire; Fire, Water; mutually, as it were, cherish one another; and by a certain unanimous consent, conspire to the Conservation of the Geocosm, or Terrestrial World.'[8] Put more simply, volcanoes acted as safety valves to release underground combustion and to prevent the Earth from overheating. In this sense, volcanoes are construed primarily in architectural terms as 'ducts', 'funnels', 'chimneys', as 'nothing but the vent-holes and breath-pipes of Nature'. They are conceived of as deep structural features and, as we have seen, are rendered in the classical

6. Cocco, *Watching Vesuvius*, p.154-55.
7. This is most famously illustrated in the figure 'Systema Ideale Pyrophylaciorum Subterraneorum, quorum montis Vulcanii, veluti spiracula quaedam existant', in Kircher, *Mundus subterraneus*, p.180-81.
8. *The Vulcano's, or Burning and fire-vomiting mountains, famous in the world* (London, J. Darby, 1669), p.ii.

form of analogies. As such, they were easy for contemporaries to grasp; something which prolonged the influence of Kircher's holistic theory well into the eighteenth century.

The Italian priest Giacinto Gimma adopted Kircher's vision of volcanoes as the surface openings of criss-crossing subterranean channels and tunnels running with rivers of fire in his *Fisica sotterranea* of 1730.[9] The same theory was given a providentialist twist in the abbé Noël-Antoine Pluche's bestselling *Spectacle de la nature* (1732-1742) in which a beneficent deity opened up volcanoes 'at intervals' across the Earth's surface to act as 'spiracles through which all the [underground] air escapes along with everything that has been set ablaze in the bowels of the Earth'.[10] In turn, the baron Paul-Henri Thiry d'Holbach's article 'Volcans' in the *Encyclopédie* was to reprise this image of volcanoes as 'vents' or 'chimneys' through which the Earth 'rids itself of the burning matter consuming its insides'.[11] Kircher's theory of intersecting channels of subterranean fire and water was also put forward to explain the terrible Lisbon earthquake of November 1755 in which as many as 30,000 people lost their lives to a fatal combination of seismic shock, sweeping tsunami and city fires.[12]

Yet already in the mid-seventeenth century Kircher's abstract, global theorising about volcanoes and earthquakes had been challenged by alternative interpretations of volcanic activity. One of the more coherent conceptions was that of Giovanni Alfonso Borelli, professor of mathematics at the University of Messina in Sicily. His understanding of volcanism was shaped by his eyewitness experience of the mighty Etna eruption of March 1669. In his *Historia et meteorologia incendii Aetnaei anni 1669*, published in 1670, Borelli thus privileges localised and empirical evidence as the basis of his ideas on volcanoes. Specifically, he studied the mechanics of the huge, swelling lava flows pouring from Etna and drew on a chemical model to conceptualise their combustion within the volcano. Essentially, an eruption was occasioned when the confined fermentation of seams

9. Giacinto Gimma, *Della storia naturale della gemme, delle pietre, e di tutti i minerali, ovvero Della fisica sotterranea*, 2 vols (Naples, Gennaro Muzio, 1730), vol.2, p.493-99.

10. Noël-Antoine Pluche, *Le Spectacle de la nature, ou Entretiens sur les particularités de l'histoire naturelle*, 9 vols (Paris, veuve Estienne, 1732-1742), vol.3 (1735), p.269-70.

11. Paul-Henri Thiry, baron d'Holbach, 'Volcans', in *Encyclopédie, ou Dictionnaire raisonné des sciences, des arts et des métiers*, ed. Denis Diderot and Jean D'Alembert, 35 vols (Paris, Briasson, 1751-1780), vol.17 (1765), p.443-45 (443).

12. See the excellent study by Grégory Quenet, *Les Tremblements de terre aux XVIIe et XVIIIe siècles: la naissance d'un risque* (Seyssel, 2005), p.451.

of sulphur and bitumen in the mountain sought explosive release; the torrents of lava thus produced were composed of a molten layer of surface rock and – despite Borelli possessing evidence to the contrary – substantial amounts of liquid sulphur and bitumen. This mechanist-chemical model of local volcanism meant that he rejected outright Kircher's speculative theory of a central core of fire in the Earth; and his emphasis on surface proofs induced him to propose a much shallower seat for eruptions. In this much, Borelli's thinking on volcanism eschewed the macro-structural approach of *Mundus subterraneus*; and, if it retained an analogical model, it was now a chemical one, specifically the concussive and explosive force of loosely packed gunpowder.[13]

Borelli's ideas found a receptive audience in the new state academies of science in France and England. As a correspondent of the Royal Society in London, the Italian brought first-hand testimony of an eruption to a scientific debate on volcanism that had already taken a chemical turn. Robert Hooke's occasional lectures on earthquakes, given in the same institution from 1668 to 1699, similarly proposed an analogical model with gunpowder, conflating thunder, lightning and volcanic eruptions as instances of the explosive ignition of nitrous and sulphurous vapours confined either in clouds or in the earth.[14] Yet it was another member of the Royal Society, the British physician and naturalist Martin Lister, who developed and popularised this chemical explanation for volcanic activity in the early 1680s. In three successive papers published in 1684, Lister singled out pyrite (otherwise known as fool's gold) as the active agent in thunder and lightning, earthquakes and volcanic eruptions. As in the thinking of Hooke and Borelli, what initially linked pyrite to volcanoes was the presence of sulphur smelled in the air and seen in the yellow brimstone formed on the rim of craters and fumeroles. Sulphur, long known as a primary chemical substance, was a major component of pyrite. In fact, pyrite, according to Lister, generated heat, even fire, precisely as its constituent elements of sulphur and iron fermented, like a hellish yeast, to the point of self-ignition. Thus he writes: 'we,

13. Borelli, *Historia*. On Borelli's specific conceptions of volcanism, see Cocco, *Watching Vesuvius*, p.156-69.
14. Hooke's lectures were only published posthumously in *The Posthumous works of Robert Hooke* (London, Sam. Smith and Benj. Walford, 1705), 'Lectures and discourses of earthquakes, and subterranean eruption', p.265-450, see, for example, p.416-26.

with great probability, believe [volcanoes] to be Mountains made up in great part of Pyrites, by the quantities of Sulphur sublimed, and the Application of the Load-stone to the ejected Cinder [i.e. their high iron content].'[15]

Lister's theory was, in turn, enthusiastically espoused by the French empirical chemist Nicolas Lémery. In a series of experiments, Lémery mixed and buried quantities of sulphur, nitrate and iron filings in earth and fed them with varying amounts of air and water to produce small subterranean combustions which, he claimed, replicated on a miniature scale the chemistry of volcanic activity. In the tenth edition of his popular *Cours de chymie* (1713), Lémery repeated the assertion that it was the intense friction of sulphur and iron particles which created heat and fire. This process of chemical fermentation, triggered in huge seams trapped beneath Vesuvius and Etna, was therefore what caused the volcanoes to erupt. He pointed to large quantities of iron-rich scoriae on Etna which resembled iron slag burned off in forges as empirical proof that the same processes were occurring on the volcano as were carried out in his experiments.[16]

The great advantage of Lémery's application of Lister's theory was that it could be endlessly reproduced in the *cabinets d'étude* or private laboratories of amateur scientists and gentleman chemists across Europe. In its use of basic chemical materials – iron, sulphur, nitrate, air, water – it was also free of the arcane nomenclatures of alchemy at a time when modern chemistry had yet to emerge as a discrete empirical scientific discipline. Hence we find Lémery's chemical demonstrations of volcanism complementing, and ultimately eclipsing, Kircher's theory in the article 'Volcan' of the *Encyclopédie.* (This article, written by the chevalier Louis de Jaucourt, follows directly on from d'Holbach's remarks cited above on 'Volcans' in the plural.) De Jaucourt notes: 'Earthquakes and volcanoes depend on the same cause, namely, earth which contains a lot of sulphur and nitrate being set on fire by the inflammable vapours from pyrite [...]

15. Martin Lister, 'Three papers of Dr Martin Lyster, the first of the nature of earth-quakes, more particularly of the origine of the matter of them, from the pyrites alone'; 'The second paper of the same person concerning the spontaneous firing of the pyrites'; 'The third paper of the same person, concerning thunder and lightning being from the pyrites', *Philosophical transactions* 14 (1684), p.512-19 (516). See also the interesting article by Anna Marie Roos, 'Martin Lister (1639-1712) and fools' gold', *Ambix* 51:1 (2004), p.23-41.
16. Nicolas Lémery, *Cours de chymie*, 10th edn (Paris, Jean-Baptiste Delespine, 1713), p.179.

volcanoes are just so many vents which serve to release the sulphurous matter sublimated by the pyrite.'[17]

To many this conception seemed to offer a yet more compelling explanation for the Lisbon earthquake; and, indeed, it was to have its champions well into the latter half of the eighteenth century. For instance, the great French astronomer Joseph Jérôme Lalande, visiting Vesuvius in 1766, thought it most likely that eruptions occurred when seawater infiltrated the volcano and came into contact with pyrite at its base.[18] He drew this idea not just from Lémery, but also cited among his recently published sources Johann Gottlob Lehmann's mining manual *L'Art des mines* (1759) and Johann Friedrich Henkel's *Pyritologie, ou Histoire naturelle de la pyrite* (1760). And as late as 1799, the same year in which he was elected president of the Royal Irish Academy, Richard Kirwan offered a spirited defence of Lémery's chemical conception of volcanism in his *Geological essays*. Claiming that lava was less hot than was 'commonly believed', Kirwan explained that volcanic heat originates firstly in the 'decomposition of water by pyritous masses', which fuses the iron and sulphur contained in the pyrite and liquefies adjacent coal seams. The mountain's combustibles, however, are only 'fired with dreadful explosions, and immense volumes of flame', when the sulphur and bitumen vapours ignite violently on contact with air.[19]

As with Lalande, Kirwan's continuing promotion of a predominantly chemical exposition of the workings of volcanoes could not feasibly rest on Lémery's small-scale experiments alone. It was also supported, albeit indirectly, by contemporary chemists' eager involvement in the extensive fieldwork carried out in volcanic terrain across Europe during the last decades of the eighteenth century. Indeed, as Corinna Guerra has convincingly shown, the lack of institutional sites for chemical experimentation in late-eighteenth-century Naples obliged educated, local chemists, such as Giuseppe Vairo, to use the neighbouring Solfatara sulphur fields and Vesuvius itself as an open-air chemical laboratory.[20] While many of these experiments concerned the chemical agency of specific volcanic products, they necessarily also connoted a strong geochemical understanding of

17. Louis de Jaucourt, 'Volcan', in *Encyclopédie*, ed. D. Diderot and J. D'Alembert, vol.17, p.445-46 (446).
18. Lalande, *Voyage d'un François*, vol.7, p.182.
19. Richard Kirwan, *Geological essays* (London, T. Bensley, 1799), p.269-70.
20. Corinna Guerra, 'If you don't have a good laboratory, find a good volcano:

volcanism. Vairo, for instance, was a proponent of the pyrite-and-seawater conception of volcanic activity favoured by Lalande.[21] This situation was complicated somewhat by the revolution in chemical thinking effected by Antoine-Laurent Lavoisier in 1789 when he introduced a modern system of classifying basic chemical elements and their compounds. Hence the cataclysmic eruption of Vesuvius in 1794 led the eyewitness chemist Emanuele Scotti to apply Lavoisier's new terminology to the tremendous cloud-bursts produced by the roaring uprush of tephra, asserting that they were the result of hydrogen gases from combusted pyrite reacting with oxygen in the atmosphere to produce torrential rains. Scotti, nonetheless, felt obliged to temper this highly modern interpretation by suggesting a seawater-and-pyrite reaction had also contributed significantly to the sheer volume of volcanogenic thunderstorms unleashed.[22]

As far as specifically chemical conceptions of volcanism are concerned, it is finally worth noting that, around the same time that Vairo and Scotti were carrying out *in situ* analyses of volcanic chemical reactions, Lémery's original experimentation appeared in a much less likely context, deployed perversely in the marquis de Sade's pornographic novel *La Nouvelle Justine* (1797). In a Sicilian interlude, Sade's debauched monk Jérôme teams up with the libertine chemist Almani to execute a dastardly plot to blow up the eastern side of the island. Their evil plan involves planting thousands of 'bombs' containing an explosive mixture of water, iron powder and sulphur in a twenty-mile circle around the base of Etna. The volcano's hot soils ferment and ultimately ignite this deadly cocktail, causing 'one of the most terrible upheavals to shake the island for several centuries: 10,000 houses were knocked over in Messina, five public buildings flattened, and 25,000 souls fell prey to our [Jérôme and Almani's] signal wickedness'.[23] In his criminal imagination, Sade conflates Lémery's experiments with the devastating Calabrian earthquake of

---

Mount Vesuvius as a natural chemical laboratory in eighteenth-century Italy', *Ambix* 62:3 (August 2015), p.245-65.

21. See Vairo's translation of, and annotations to, the first two volumes of Pierre Joseph Macquer's chemical dictionary, *Dizionario di chimica*, 10 vols (Naples, G.-M. Porcelli, 1784-1786), vol.1, p.328. Cited in Guerra, 'If you don't have a good laboratory', p.257.
22. Emanuele Scotti, *Della eruzione del Vesuvio, accaduta il dì 15 giugno 1794* (Naples, Raffaele Lanciano, 1794), p.13, 29. Cited in Guerra, 'If you don't have a good laboratory', p.262.
23. Sade, *La Nouvelle Justine*, vol.2, p.45.

February 1783 which wrought the havoc described in the novel on both sides of the Strait of Messina. Human agency and geological accident are made to conspire, since for Sade both are the fruits of a fundamentally indifferent and immoral nature.

Volcanological theories based on seawaters and electricity

The source for Sade's depraved exploitation of Lémery's science could have been the *Encyclopédie*. It could equally well have been one of the most influential of all scientific writings on volcanoes in the eighteenth century, Georges-Louis Leclerc de Buffon's seminal *Théorie de la Terre* of 1749, published as part of the first volume of his exhaustive *Histoire naturelle*. Buffon was called the 'modern Pliny' and his *Théorie* served as the scientific primer for many grand tourists visiting Vesuvius or Etna later in the century. Carrying the seal of approval of the top academies in France, Buffon's work was diffused across Europe.[24] It reached beyond institutional holdings and was found on the shelves of private studies and in public libraries not just in France but abroad too. For instance, Roy Porter reports it being held by Bristol Library in the 1770s and even by Philpott's circulating library operating in the same city, sitting alongside a copy of Rudolf Erich Raspe's *An Account of some German volcanos, and their productions* (1776).[25]

Buffon's debt to Lémery is clear from his opening presentation of how a volcano erupts:

> Veins of sulphur, of bitumen and other inflammable matter are to be found in a mountain, at the same time there are found other minerals, pyrite which may ferment, and does ferment each time that it is exposed to air or to moisture, all of these are to be found together in very large amounts, fire is kindled and causes an explosion proportionate to the quantities of burning matter present [...]

It is even more evident in what follows:

> This is what a volcano is for a natural scientist, and it is easy for him to imitate the action of these subterranean fires by mixing together

24. See Grell, *Herculanum et Pompéi*, p.23, 168-73; for an example of Buffon as 'le Pline moderne', see Nicolas Bricaire de La Dixmerie, *Les Deux âges du goût et du génie français, sous Louis XIV & sous Louis XV* (Amsterdam, Barthelemi Vlam, 1770), p.91.
25. Roy Porter, *The Making of geology: earth science in Britain 1660-1815* (Cambridge, 1977), p.99.

> a certain quantity of sulphur and iron filings which are then buried at a certain depth and thus make a miniature volcano the effects of which are the same, relatively speaking, as those of large volcanoes, it catches fire by fermentation alone, throwing up the earth and stones covering it, and makes smoke, flames and explosions.[26]

He goes on to list ancient and more recent eruptions around the globe, seamlessly moving on to a similar list of major earthquakes. Yet, having spelled out the universal devastation wrought by volcanoes and seismic shocks throughout history, he diminishes their importance as geological forces shaping the face of the Earth. His argument turns on the fact that volcanoes are not composed of successively deposited strata like all other mountains; their 'small elevations' are thus the exception, not the rule, as far as orogeny or mountain building is concerned. Are not marine fossils further proof that mountains are produced not by irregular, localised bursts of volcanic activity, but by the more regular, widespread, long-term actions of the seas?[27] Buffon concedes that volcanoes may make powerful impressions on the local landscape, blasting out enormous craters and caverns or pouring forth extensive lava fields. In this much, they contribute to the subterranean confusion of rocks, minerals and metals and partake of the apparent surface disorder of the planet.[28] But this is the limit of their world-sculpting powers. Little or nothing permanent results from their discontinuous, catastrophic explosions. For Buffon, it is not the 'rare, violent or sudden' geological phenomena which shape the Earth, but sustained, repeated, regular and continuous forces – specifically those of water, that is, the processes of sedimentation and erosion.[29]

Buffon's vision of a world made by the actions of sweeping bodies of water draws on the prevailing theories of the so-called 'Neptunist' school. Dismissing the more deep-earth theories of volcanism and other geophysical forces, scholars such as Louis Bourguet and Benoît de Maillet saw topographies as being moulded by continuous, recurrent physical forces operating in the world around them.[30] Chief among

26. Georges-Louis Leclerc, comte de Buffon, *Histoire naturelle, générale et particulière*, 36 vols (Paris, Imprimerie royale, 1749-1788), vol.1, p.503-504.
27. Buffon, *Histoire naturelle*, vol.1, p.77-82.
28. Buffon, *Histoire naturelle*, vol.1, p.114-15, 68-69.
29. Buffon, *Histoire naturelle*, vol.1, p.99, 523-24.
30. See Oldroyd, *Thinking about the Earth*, p.88-89; see Louis Bourguet, *Traité de pétrifications* (Paris, Briasson, 1742), and Benoît de Maillet, *Telliamed, ou Entretiens*

these was water. De Maillet's posthumous *Telliamed* (1748) – the title of which is but his name spelled backwards – thus contains wildly speculative visions of a global, primordial sea acting on the Earth over millions of years.[31] This 'universal ocean' and its vast geological timeframe would both be notions that Buffon would develop, albeit more circumspectly and systematically. But this economy of natural causes, privileging at every turn water over fire, necessarily relegated volcanism to the status of an incidental phenomenon in the *Théorie de la Terre*. Benoît de Baere has argued that in such a self-conscious stylist as Buffon the pervasive image of the sea also has its part to play in determining his theory. Figuratively as well as theoretically, the tidal trumps the volcanic. In other words, favouring continuous cycles of water-powered change gives continuity to his geological narrative, sequencing its layers of assertion and deduction, and eroding the arguments of its opponents. Violent, sporadic volcanic action does not fit into Buffon's conceptual framework because the perceived order of nature is modelled in part on the progressive order of narration.[32]

Three years after the publication of Buffon's *Théorie*, the celebrated cartographer and royal geographer Philippe Buache endorsed Buffon's vision of volcanoes as mountain anomalies in his highly influential *Essai de géographie physique*. Buache pared down his conception of the world to two interlocking systems: one of uninterrupted mountain ranges that 'cross and gird' the Earth, acting as its chassis or frame, and another of water basins and water courses. As in Buffon, water is paramount: seas and rivers demarcate the mountain ranges whose chief function is to divide and direct the waters. Volcanoes may pepper these mountain chains as they rise over continents or run under oceans, but they are very much secondary features.[33] Once

d'un philosophe indien avec un missionnaire françois sur la diminution de la mer, la formation de la terre, l'origine de l'homme, &c., 2 vols (Amsterdam, L'Honoré & Fils, 1748).

31. See Albert V. Carozzi, 'De Maillet's *Telliamed* (1748): an ultra-Neptunian theory of the Earth', in *Toward a history of geology*, ed. Cecil J. Schneer (Cambridge, MA, 1969), p.80-99.
32. Benoît de Baere, 'Natural catastrophe in Buffon's *Histoire naturelle*: earth, science, aesthetics, anthropology', in *Histoires de la Terre: earth sciences and French culture 1740-1940*, ed. Louise Lyle and David McCallam (Amsterdam, 2008), p.19-35.
33. Philippe Buache, 'Essai de géographie physique, où l'on propose des vûes générales sur l'espèce de charpente du globe, composée des chaînes de montagnes qui traversent les mers comme les terres, avec quelques considérations particulières sur les différens bassins de la mer, & sur sa configuration intérieure', *Mémoires de l'Académie royale des sciences* (1752), p.399-416.

again, volcanism finds no place in a system built upon notions of continuity and evolutionary change; a point belatedly raised in criticism by the baron Paul-Henri Thiry d'Holbach in his article 'Montagnes' in the *Encyclopédie*.[34] In effect, the figure of the mountain 'chain' determines, indeed overdetermines, the theory it is only supposed to illustrate, to the extent that its continuous links induce the geographer to project mountain ranges across plateaux and under seas where in reality none exists.[35] Notwithstanding these flaws, Buache's 'armchair' geography was to remain popular well into the early nineteenth century, institutionally sponsored, easy to understand and simple to teach.

Buffon's influence was not, however, altogether deleterious for an appreciation of volcanoes in the eighteenth century. His *Théorie* did class volcanoes among the effective causes of geological and topographical change, even if they belonged to the less significant category of 'particular' causes, rather than to that of 'general' ones.[36] That is, they were neither vestiges of prehistoric chaos nor the persistent object of superstitious fear. The ignorant populace might still mistake them for the 'mouths of hell' or perpetuate the superstitious beliefs of medieval monks that volcanic rumblings amount to the tortured cries of the damned;[37] more enlightened thinkers understood that volcanism constituted an observable geological process active in the present across the world in precisely the same way as it had been in the past. This is Buffon's dictum that 'in order to understand what has occurred, and even what will occur, we have only to examine what is occurring.'[38] This so-called 'actualism' would subsequently encourage some significant fieldwork into volcanoes, even if Buffon himself did not undertake any of note.[39]

34. Paul-Henri Thiry, baron d'Holbach, 'Montagnes', in *Encyclopédie*, ed. D. Diderot and J. D'Alembert, vol.10 (1765), p.672-76.
35. See the important study by Numa Broc, *Les Montagnes au siècle des Lumières, perception et représentation* (Paris, 1991), especially p.56-70; also David McCallam, 'Quelques réflexions sur la figure de la chaîne au XVIIIe siècle: en théorie et en pratique', in *De la théorie à l'action: les savoirs et leur mise en œuvre au siècle des Lumières*, ed. Valérie Kobi (Geneva, 2011), p.123-34.
36. Buffon, *Histoire naturelle*, vol.1, p.609.
37. See chapters 5 and 6 in the present volume.
38. Buffon, *Histoire naturelle*, vol.1, p.96.
39. For an interesting short study problematising this conception of 'actualism', see Reijer Hooykaas, *Catastrophism in geology, its scientific character in relation to actualism and uniformitarianism* (Amsterdam, 1970), on Buffon p.10-13.

Moreover, from the mid-1760s onwards his conception of the Earth shifted in emphasis, reconfigured under the influence of work done by Jean-Jacques Dortous de Mairan on the core heat of the globe.[40] This led Buffon to conduct some experiments of his own on global heat and global cooling. Large metal spheres were subjected to heat until white-hot, then allowed to cool until, rather unscientifically, the family milkmaids on his estate at Montbard in Burgundy could 'touch them and hold them in the hand, for a second, without burning [themselves]'.[41] The spheres acted as model worlds, and their gradual cooling was believed to reproduce in scaled-down form the unilateral cooling of the Earth itself.[42] For this planetary cooling had become Buffon's new master narrative for conceptualising the geological history of the globe. It is enshrined in his *Epoques de la nature* of 1778 which recount the evolution of Earth history over seven geological periods or epochs. In brief, the new theory maintained that our planet had been formed as ejected matter when a comet struck the sun; in the first epoch this molten mass takes the shape of a spheroid; in the second, it solidifies; in the third, atmospheric vapours are precipitated and form a universal ocean 'up to 4000 metres [deux mille toises] above current sea level'; in the fourth, intense seismic and volcanic activity occurs, triggered by the Earth's internal heat reacting with waters of the 'universal sea' and sparking explosive fires in the pyrite-rich, coal-laden sediments; in the fifth, relative calm returns, with tropical life forms populating the poles, as the equatorial regions are still too hot to sustain life; in the sixth, as the globe continues to cool, flora and fauna migrate south towards the equator, and humankind appears; in the seventh, humans assert their supremacy – this epoch continues to the present, and will last until the Earth loses all its internal heat and life becomes extinct.[43]

This conceptual framework retained much of its intellectual authority until the end of the nineteenth century, when the discovery

40. Jean-Jacques Dortous de Mairan, 'Nouvelles recherches sur la cause générale du chaud en été & du froid en hiver, en tant qu'elle se lie à la chaleur interne & permanente de la Terre', *Histoire de l'Académie royale des sciences, année M. DCCLXV* (Paris, Imprimerie royale, 1768), p.143-267.
41. Buffon, *Histoire naturelle: supplément*, vol.1 (Paris, Imprimerie royale, 1774), p.147.
42. See Buffon, *Histoire naturelle: supplément*, vol.2 (Paris, Imprimerie royale, 1775), 'Premier mémoire: recherches sur le refroidissement de la Terre et des planètes', p.361-515.
43. Buffon, *Histoire naturelle: supplément*, vol.5: *Des époques de la nature* (Paris, Imprimerie royale, 1778).

of radioactivity suggested that alternative forms of energy generation might also be fuelling core heat, so making a nonsense of Buffon's extensive calculations.[44] (The *Epoques* would jar yet more gratingly with current ecological sensibilities, insofar as they unashamedly urged deforestation, wetland draining and population increase as intelligent, immediate ways of retarding, if not checking, the irreversible loss of the Earth's internal heat. Clearing, drying and peopling the planet are, Buffon argues, welcome means of raising its falling surface temperature.)[45]

One advance of Buffon's new theory, driven by the processes of global cooling, is that natural catastrophes find their place in his overall schema of Earth history. Thus earthquakes and volcanic eruptions proliferate violently in his fourth epoch; although the implication is that in subsequent periods their activity gradually lessens and may eventually die out altogether. Volcanoes have acquired a temporally specific geophysical dynamic – that of core heat combining with seawater to set off massive conflagrations in pyrite-rich coal seams – but, at the same time, they remain secondary telluric forces, flaring up impressively in the fourth epoch but guttering gradually thereafter.[46] What is more, they remain subordinate to the catalysing power of water. Buffon states that 'pyrite-rich combustible matter [...] requires a certain amount of water in order to effervesce, and it is only then, as a result of a great volume of fire colliding with a great volume of water, that violent eruptions can occur.' In theoretical terms, Buffon sets a Kircherian detonator (clashing masses of fire and water) into a seam of Lémerian explosives (pyrite). It also follows from his understanding of volcanism that 'all currently active volcanoes are situated on islands or near the sea'; that just as 'there existed no active volcanoes before the seas were formed', so volcanoes will only remain active as long as the seas do not recede and deprive them of their saltwater primers.[47] Buffon imagines here that volcanic combustion takes place at a shallow subterranean level, kindling in the iron and sulphur-rich deposits directly beneath or beside the vent. He shares in this a contemporary ignorance of the vast boiling magma chambers situated at a depth of several miles below the crater. Ironically, however, recent

44. Barbara A. Kennedy, *Inventing the Earth: ideas on landscape development since 1740* (Oxford, 2006), p.1, 11.
45. Buffon, *Epoques de la nature*, p.240.
46. Buffon, *Epoques de la nature*, p.131-64.
47. Buffon, *Epoques de la nature*, p.134.

geological research has proved Buffon half-right. His pyrite and coal explosions are clearly fanciful, but in the subduction zones where we now know tectonic plates collide and subsume one another, the most volatile and unstable magma reserves are formed by the subduction of large volumes of seawater and sea-floor sediments along with the more solid crust. These sea-fed magmas are viscous, gas-filled masses, most apt to force a path to the surface via volcanic vents.[48] Seawater therefore is responsible in some measure for explosive eruptions, but only as part of a 'deeper' truth to Buffon's 'superficial' hypothesis on volcanism.

Yet the *Epoques* are not content to conscript Kircher and Lémery into their explanations of volcanoes. They also draw on newer, more dazzling discoveries to bolster their theory. Hence, writes Buffon, 'electricity seems to me to play a very large role in earthquakes and volcanic eruptions'.[49] However, electricity is not arbitrarily co-opted into a list of potential causes of volcanism. On the contrary, Buffon here effects a canny synthesis of electrical theories and his own experiments on global heat. For, he maintains, electricity is in fact nothing other than an invisible emanation of the internal heat of the globe, invisible, that is, until it is stifled or checked when it manifests itself as sparks, shocks, currents or lightning. Imagine then that this electrical power is trapped by the mass and solidity of the Earth, would not this 'subterranean lightning' trigger seismic tremors and volcanic eruptions?[50] Nonetheless, Buffon is careful not to forget his initial premise concerning volcanism; it is not that these underground electrical storms supersede pyrite-based explosions as the cause of eruptions, rather they 'combine' with them to increase the violence of volcanic activity. Unleashed together these chemical and electrical forces thus account for all the 'principal phenomena' associated with volcanoes.[51] Exploring the Dalmatian hinterlands in the early 1770s, the Italian geologist Alberto Fortis seems to give some credence to this idea. Addressing his patron the earl of Bute, Fortis remarks that 'among the original causes of volcanic mountains we

48. Clive Oppenheimer, *Eruptions that shook the world* (Cambridge, 2011), p.14-15.
49. Buffon, *Epoques de la nature*, p.136.
50. Buffon, *Epoques de la nature*, p.136-37. This was equally a classical idea: in book 8 of the Virgil's *Aeneid* Venus asks Vulcan to forge Aeneas' armour. The fire god goes to his smithies under Vulcano to set his Cyclops to work and finds them there labouring iron to fashion Jove's thunderbolts.
51. Buffon, *Epoques de la nature*, p.137.

have a right also to place lightning.' His reflection is inspired by the anecdote of a peat bog set ablaze by a lightning strike and burning for some time thereafter. Fortis then extrapolates from this freak occurrence that 'if a thunderbolt were to fall in a mountain of sulphur, would it not in all probability make a much greater noise and have more remarkable consequences than falling on the wet peatlands of Ostrovizza [Ostrovica]?'[52]

Of course, electricity had long been one of the star attractions of the European scientific community. From the invention of the Leyden jar in the 1740s to store static electricity, through the amazing publicity surrounding the electrical experiments of the abbé Jean-Antoine Nollet and Benjamin Franklin in the 1750s, this natural force had been credited with the power to do everything from curing disease to explaining sexual reproduction.[53] In the earth sciences, electricity proved the most popular theory for comprehending the cause and effects of the Lisbon earthquake. It was variously championed in academic essay-writing competitions on the subject, in scientific journals and in debates within learned societies.[54] What appealed most about electricity as a potential explanation for earthquakes – and, by extension, volcanic eruptions – was its incredible energy and speed. In the 1740s experiments conducted across vast bodies of water at Lake Geneva and in London demonstrated that electricity could travel at up to thirty times the speed of sound without any significant diminution of its power.[55] To contemporaries this explained why the Lisbon tremors were felt almost simultaneously and equally strongly in London, Rome and Switzerland as in Spain.

The most zealous advocate of electricity in France, the abbé Pierre-Nicolas Bertholon had no hesitation in claiming that what he called 'electrical fluid' was the 'sole cause of a great number of meteorological phenomena', including seismic shocks and volcanic eruptions. In *De l'électricité des météores* (1787) he asserts the validity of the electrical

52. Alberto Fortis, *Viaggio in Dalmazia*, 2 vols (Venice, Alvise Milocco, 1774), vol.1, p.39.
53. See François Zanetti, *L'Electricité médicale dans la France des Lumières* (Oxford, 2017).
54. Quenet, *Les Tremblements de terre*, p.368. See also the same author's 'When geology encounters a real catastrophe: from theoretical earthquakes to the Lisbon disaster', in *Histoires de la Terre*, ed. L. Lyle and D. McCallam, p.37-56 (49-50).
55. John Lewis Heilbron, *Electricity in the 17th and 18th centuries: a study of early modern physics* (Berkeley, CA, 1979), p.320.

theory of volcanism to the exclusion of all other explanations – be they 'underground winds, subterranean torrents, inflammable emissions, pockets of steam'. For Bertholon, electricity alone is able to account for the instantaneous, powerful and inexhaustible nature of volcanic activity: it is unimaginably fast, as has been proven, its power is incontrovertibly manifest in the effects of lightning, a phenomenon which also illustrates its continuously renewable quality. This last point is important, since pyrite deposits and coal seams are finite, whereas the ceaseless turmoil of a volcano such as Stromboli requires an equally limitless energy source.[56]

If, in Bertholon, we find the same 'subterranean thunder and lightning' evoked as a conceptual model of volcanism as we had found in Buffon's *Epoques*, Bertholon nonetheless takes the analogy much further.[57] In fact, he derives some astounding practical consequences from it. If volcanoes can be perceived as one of nature's more radical ways of venting a build-up of 'electrical fluid' within the Earth's interior, it should then be possible to diffuse the threat posed by volcanoes (and earthquakes) by discharging the excess electricity harmlessly into the atmosphere. This can most obviously be achieved by inverting the mechanics of Franklin's lightning conductor, or 'paratonnerre'. Thus, in a paper published in the *Journal encyclopédique* of 1779, Bertholon proposes what he calls the 'para-volcano' or volcano conductor. This is a sort of enormous upside-down lightning rod consisting of large metal stems sunk deep into the earth ending in fine, splayed points called 'verticils'. These are specifically designed, writes Bertholon, to draw the pent-up electricity out of the earth and channel it safely into the air. He goes on to suggest that trials of a prototype might be most usefully carried out near Vesuvius or Etna.[58] As far as we know, however, his para-volcano never got beyond the design stage.

What Bertholon and Buffon share in their writings on volcanoes is a desire to subordinate them to a catch-all geological theory, to make them only further illustrations of the all-embracing power of either the seas or electricity. Observation is sacrificed ultimately to theoretical coherence; something that is all the easier to do insofar as

56. Pierre-Nicolas Bertholon, *De l'électricité des météores*, 2 vols (Paris, Croullebois, 1787), vol.1, p.x, 355, 357-59.
57. Bertholon, *De l'électricité des météores*, vol.1, p.344.
58. Pierre-Nicolas Bertholon, 'Mémoires sur un para-tremblement de terre et un para-volcan', *Journal encyclopédique* 3 (1779), p.123-32, 323-31. See also Quenet, *Les Tremblements de terre*, p.445-49.

neither Bertholon nor Buffon had seen an active volcano first-hand. Buffon strolled and botanised on the volcanic plateau of Langres, and Bertholon had examined the extinct volcanoes of Languedoc. But, in their writings, they cite others' testimonies of volcanic eruptions: for instance, in the case of Bertholon, Sir William Hamilton on Vesuvius or Déodat de Dolomieu on the Aeolian Islands, or, further afield, James Cook and Georg Forster on Tanna island in the South Pacific archipelago.[59] The abstract Volcano, as opposed to physical volcanoes, becomes a space of theoretical speculation and free-ranging fancy all the more to be indulged because this space is subterranean and unfalsifiable. Hence both authors more or less close their eyes and take the imaginative leap with Empedocles into the crater. And in the conceptual dark of the volcano's interior, it is not sight but other senses, especially hearing and smell, which guide these seekers after volcanic truths. Buffon and Bertholon both seize on the thunder of the volcano, and accordingly construct an image of it that privileges its rumbling, bellowing, howling noise. This 'underground thunder' supposes lightning; lightning supposes electricity; and what started as a meteorological figure (thunder) ends up confirming its theoretical premise (electricity).

Yet the logical implications of this imaginative and rhetorical straining to comprehend the volcano do not stop there. In Buffon's *Epoques*, for example, volcanic eruptions are part of a larger poetics of natural violence. They number among the more visible and spectacular proofs of nature's profound indifference to mankind. Here then is Buffon's primitive humanity; it stands 'witness to the still recent and very frequent convulsions of the Earth, having only mountains for refuge against floods, often chased from these places of refuge by volcanic fire, quaking on earth that quaked beneath their feet, naked in mind and body, exposed to the fury of the elements, victims of ferocious wild animals'.[60] We could not be farther from Rousseau's self-sufficient hominids or his free, happy and healthy hunter-gatherers.[61] And as this volcanic fire threatens mankind, so mankind must unite to master it. As European societies subsequently rise from their terrified, superstitious beginnings, Buffon suggests

59. Bertholon, *De l'électricité des météores*, vol.1, p.313, 357, 307 respectively.
60. Buffon, *Epoques de la nature*, p.225.
61. See Jean-Jacques Rousseau, *Discours sur l'origine et les fondemens de l'inégalité parmi les hommes*, in *Collection complète des œuvres*, 17 vols (Geneva, Société typographique de Genève; Barde, Manget & Cie, 1780-1789), vol.1, p.1-176.

that the civilising process will naturally be accompanied by the gradual extinction of the volcanism that so harried and horrified their ancestors. What is more, in keeping with the progressive thinking behind Bertholon's para-volcanoes, eighteenth-century man – and it is a male world – can do more than simply watch volcanoes peter out in their own slow time: he can, claims the *Epoques*, accelerate this natural process by cutting off the supplies of seawater to volcanoes, thereby depriving them of the principal trigger of their eruptions. This should not cost more effort, Buffon maintains, than it did to build the 'vainglorious pyramids of Egypt'. It would employ the same forces but construct 'monuments to our wisdom', as we might thus at last 'master Nature'.[62] These are the dizzy heights to which abstract volcanological theorising led not just Bertholon and Buffon, but also the Italian naturalist Scipione Breislak as he contemplated Vesuvius in the 1790s, stating boldly: 'Do not let ourselves be dazzled by the majesty of nature, there is nothing so great and so powerful that man cannot master it.'[63] A positivist confidence in the powers of science over nature that is of a piece with the risible injunction of Frederick II in the wake of the Lisbon disaster. For the Prussian monarch, earthquakes were quite simply outlawed in his territories.[64]

The basalt controversy and the empirical turn

Far from sating the contemporary fascination for all things volcanic, the speculative geological theories of the likes of Buffon and Bertholon only fuelled a debate that would intensify dramatically from the 1760s to the French Revolution. According to Grégory Quenet, from 1765 onwards, the registers of the Académie royale des sciences show that a fascination with volcanism rivals then outstrips the intense earlier interest in earthquakes expressed in the immediate wake of the Lisbon disaster. To give a further telling statistic: there appear three times the number of articles relating to volcanoes compared to seismic activity in the *Journal encyclopédique* between 1768 and 1794.[65]

62. Buffon, *Epoques de la nature*, p.140.
63. Scipione Breislak, *Voyages physiques et lythologiques dans la Campanie*, 2 vols (Paris, 1801), vol.1, p.298.
64. Cited in Quenet, *Les Tremblements de terre*, p.7-8.
65. Quenet, *Les Tremblements de terre*, p.79, 469. See also Yasmine Marcil, 'Découvrir, comprendre, ressentir la montagne dans la presse périodique des années 1780', *Compar(a)ison: an international journal of comparative literature* 1-2 (2001), p.145-70 (147-49).

Similarly, from 1767 to 1795 the Royal Society in London published Sir William Hamilton's detailed, on-the-spot reports of the lively eruptions of Vesuvius and Etna as well as related correspondence from other amateur volcanologists, such as Giuseppe Gioeni.[66] Interest was not confined to Europe either, as the first issue of the journal of the American Philosophical Society in 1771 also contained an anonymous account of the spectacular eruption of Vesuvius in October 1767.[67] The latter part of the eighteenth century thus witnessed the inception of volcanology as both a theoretical science and a field science in its own right.

Of course, the impressive cycle of major eruptions of Vesuvius (1751, 1754, 1767, 1779, 1794), Etna (1763, 1787) and the Icelandic volcanoes Katla (1755), Hekla (1766) and Laki (1783) fed the fascination and encouraged the investigation of volcanoes in Europe. Yet, prior to this intense activity, the study of volcanism had been stimulated in the early eighteenth century by the unexpected apparition of a couple of new volcanic islands off the coast of Greece and Africa respectively. The more famous of these islands emerged suddenly near Santorini in the Aegean Sea in May 1707 and its appearance was witnessed by the Jesuit Father Goree who subsequently reported it to the Royal Society in a paper published in 1711.[68] The swelling rise of the sea floor was accompanied by foul smells and discoloured waters; and it was not until almost two months after the volcanic mass surfaced that it erupted with 'fire and smoak and great noise' on 16 July 1707.[69] A further volcanic island surfaced overnight in December 1720 near

66. Hamilton's early important volcanological texts written for the Royal Society were collected in his *Observations on Mount Vesuvius* (1772); they were followed by his *Supplement to the 'Campi Phlegraei'* (1779); and the articles on 'The present state of Mount Vesuvius, with an account of a journey into the province of Abruzzo, and a voyage to the island of Ponza', *Philosophical transactions* 76 (1786), p.365-81, and 'An account of the late eruption of Mount Vesuvius [June 1794]', *Philosophical transactions* 85 (1795), p.73-116. On Gioeni's letter to Hamilton about Etna, translated by the latter, see *Philosophical transactions* 72 (1782), p.1-7. For more generally on the Royal Society's relations with southern Italy and Sicily in the early modern period, see Manuela D'Amore, *The Royal Society and the discovery of the Two Sicilies: southern routes in the Grand Tour* (Basingstoke, 2017).
67. 'An account of the eruption of Mount Vesuvius in 1767', p.281-85.
68. Father Goree, 'A relation of a new island which was raised up from the bottom of the sea on the 23rd of May 1707', *Philosophical transactions* 27 (1711), p.354-75. For more on Santorini's volcanism, see David M. Pyle, *Volcanoes: encounters through the ages* (Oxford, 2017), p.190-99.
69. Goree, 'A relation of a new island', p.360.

St Michel in the Azores; it is mentioned in the *Enyclopédie* as proof of much more widespread submarine volcanism.[70] However, it was the eruption off Santorini that inspired the Italian Anton Lazzaro Moro to elaborate his original theory of mountain building in which, unlike Buffon or Bertholon, he accords pride of place to volcanoes. His study *De' crostacei e degli altri marini corpi che si truovano su' monti* (1740) suggests that there are essentially only two kinds of mountain: 'primary' mountains welling up from the sea bottom in the form of submarine volcanoes and composed of massive unstratified rock; and 'secondary' mountains formed of the eroded deposits of successive lava flows pouring from the 'primary' mountains, consolidated in strata over vast swathes of geological time. Essentially, for Moro, all mountains are thus the product of volcanism, and the eruption of the island off Santorini is a rare but timely reminder of this fundamental geological process.[71] Nonetheless, despite his spirited promotion of volcanoes as Earth-forming forces, Moro shared with Buffon and Bertholon, among early geologists, a tendency to theorise, generalise and abstract. He had never visited an active volcano either, hence his work was also to be subjected to a gradual but unremitting revision by the generation of intrepid fieldworkers who followed him.

In fact it was neither the spectacle of gushing volcanoes on the continent nor the startling emergence of volcanic islands off its shores that sparked the most important geological controversy of the mid- to late eighteenth century. That honour was reserved for the rural backwaters of central France. For it was the discoveries made among the extinct volcanoes of Auvergne that set the burgeoning geological community buzzing and divided opinion for nearly twenty years. The principal object of this controversy was the humble rock, basalt. In 1751, while mapping this region for mineralogical exploitation in the company of Chrétien Guillaume Lamoignon de Malesherbes, the academician Jean-Etienne Guettard recognised that the terrain was largely formed of very old lava flows. He also discerned the form of sixteen to seventeen ancient craters in the region around Riom, Volvic and Puy-de-Dôme; some were grass-covered, others quarried, but all were composed of a hard, vitrified rock and were often accompanied by tell-tale mounds of pumice. Yet Guettard was by conviction a 'Neptunist', persuaded by Buffon's prevailing theory that

70. D'Holbach, 'Volcans', p.445.

71. Anton Lazzaro Moro, *De' crostacei e degli altri marini corpi che si truovano su' monti* (Venice, Stefano Monti, 1740), see especially 'Libro secondo', p.211-432.

volcanism was a secondary geophysical force and that mountainous topographies were primarily formed by the long-term effects of water-driven sedimentation and erosion. He was therefore hostile to the thinking of Moro and others that 'volcanoes [are] the sole cause of mountains', stating this to be an 'opinion I hold as false'.[72] Instead, he envisaged the Auvergne landscape as comprising volcanic peaks 'sitting on' massy mountainous bases composed of 'granite and schist' – erroneously classifying both rocks as sedimentary – whose large polygonal columns were 'too regular, too uniform' to be produced by the hectic, explosive discharge of a volcano.[73] The implication was, rather, that the prismatic basalt columns found by Guettard were precipitated from a prehistoric aqueous solution. That is, basalt was ultimately the product not of a geophysical process, but of a chemical one. As for the ancient volcanic eruptions of the Auvergne craters, these were attributed to the combustion of local coal and tar deposits. Their extinction was due, in part at least, to their no longer having any underground communication with the sea, and so, unlike active island volcanoes in Europe and elsewhere, seawater no longer catalysed their explosions.[74] Despite the muddled thinking behind the volcanism of Auvergne, Guettard's 'Mémoire sur quelques montagnes de la France qui ont été des volcans' was trailblazing in both its empirical fieldwork in the Massif Central and in its important exploitation of local knowledge on volcanic topographies.[75]

In 1763 the inspector of manufactures Nicolas Desmarest, roaming over the same lava fields near Volvic, was among the first to query Guettard's conception of the local basalts. His doubts redoubled three years later when, touring the region in the company of Louis Alexandre, duc de La Rochefoucauld, and the artist Jean-Jacques de

72. Jean-Etienne Guettard, 'Mémoire sur quelques montagnes de la France qui ont été des volcans', *Mémoires de l'Académie royale des sciences* (1752), p.27-59 (47).
73. Guettard, 'Mémoire sur quelques montagnes', p.52, 46.
74. Guettard, 'Mémoire sur quelques montagnes', p.55.
75. See Guettard's references to the field knowledge of the Clermont chemist Jean-François Ozy (who would later claim to have preceded Guettard in having first perceived the volcanism of Auvergne hills) and the lava samples sent to the academician by the local doctor, M. de L'Arbre in nearby Pontgibaud. 'Mémoire sur quelques montagnes', p.37-39, 57. See also the excellent chapter by K. L. Taylor, 'Geological travellers in Auvergne, 1751-1800', in *Four centuries of geological travel: the search for knowledge on foot, bicycle, sledge and camel*, ed. Patrick N. Wyse Jackson (London, 2007), p.73-96, which contains an exhaustive chronological index of Guettard's volcano-hunting successors in the region, p.82-88.

Boissieu, he scrutinised the basalt rock formations that Boissieu was sketching. Thus when Boissieu's drawings were subsequently submitted to Diderot for inclusion in the *Encyclopédie*, Desmarest drafted extensive notes to accompany them; notes which for the first time declared his conviction that Guettard was wrong, that basalt was indeed volcanic in origin. Drawing a striking parallel between the polygonal Auvergne basalts and the celebrated basalt columns of the Giant's Causeway in Northern Ireland, Desmarest stated categorically that 'Prismatic basalt, in articulated or non-articulated columns, is matter which has been smelted in the seat of volcanoes and spilled out in the course of eruptions'.[76] He published his arguments in more developed form in the more public forum of the *Mémoires* of the Académie royale des sciences in 1774.[77] His study entitled 'Mémoire sur l'origine et la nature du basalte' also drew on comparisons with the 'modern' lava flows of southern Italy that Desmarest had witnessed in 1767. Yet by the time his study was published, the mapping of extinct volcanoes and the hunt for volcanic extrusions were fast becoming Europe-wide activities.

Locating lucrative mineral deposits was initially a strong economic incentive for this geological exploration of formerly volcanic areas; this was the original objective behind Guettard, Lavoisier and Monnet's pioneering *Atlas minéralogique* (1780) and, as an inspector of industry, Desmarest was similarly keen to exploit mineralogical finds.[78] Elsewhere, though, the fieldwork appeared more disinterested. This is the case for the local studies carried out in the 1750s and 1770s into volcanic deposits in the Euganean Hills in north-eastern Italy by Giovanni Arduino, Alberto Fortis and John Strange, Britain's envoy to the Venetian Republic.[79] Rudolf Erich Raspe, for his part, imported Desmarest's ideas about the volcanic origin of basalt into Germany

76. Nicolas Desmarest, 'Regne minéral. Sixieme collection. Planche VII. Basalte d'Auvergne', in *Encyclopédie*, ed. D. Diderot and J. D'Alembert, vol.23 (1768), p.3-4.
77. Nicolas Desmarest, 'Mémoire sur l'origine & la nature du basalte à grande colonnes polygones, déterminées par l'histoire naturelle de cette pierre, observée en Auvergne', *Mémoires de l'Académie royale des sciences* (1774), p.705-75.
78. On the *Atlas minéralogique*, see Rhoda Rappaport, 'The geological atlas of Guettard, Lavoisier and Monnet: conflicting views of the nature of geology', in *Toward a history of geology*, ed. C. J. Schneer, p.272-87.
79. The standard works on Arduino and Fortis are respectively Ezio Vaccari, *Giovanni Arduino (1714-1795): il contributo di uno scienziato veneto al dibattito settecentesco sulle scienze della Terra* (Florence, 1993), and Luca Ciancio, *Autopsie della Terra: illuminismo e geologia in Alberto Fortis (1741-1803)* (Florence, 1995).

when he discovered outcrops of this rock type near Kassel in the late 1760s.[80] Further east, in 1773, Count Franz Joseph Kinsky discovered a small extinct volcano near Cheb in Bohemia, excitedly reporting his findings to the Hungarian mineralogist Ignaz von Born, then working at the royal court in Vienna.[81] In France, Barthélemy Faujas de Saint-Fond and Jean-Louis Giraud-Soulavie both undertook further fieldwork into the volcanic topographies of the Velay and Vivarais regions, to the south-east of Auvergne, and published important studies on them.[82] To the north, Joseph Banks, on his way to explore volcanic Iceland in 1772, marvelled at the stunningly regular prismatic columns of basalt forming Fingal's Cave on Staffa in the Hebrides, his awe-filled visit to the cavern popularised in Thomas Pennant's *A Tour in Scotland* (1774).[83] His descriptions induced other amateur geologists to make the arduous trip to the Scottish isles. Thus Faujas de Saint-Fond travelled to the area in 1784 and declared the basalt columns to be of the same stamp as those in Auvergne, recognising too on his return journey via Edinburgh that its castle sat on an exposed volcanic plug, also composed of basalt.[84] In the 1760s and 1770s it seemed that

80. Rudolf Erich Raspe, 'Nachricht von einigen niederhessischen Basalten', *Schrifter der Königlicher Societät der Wissenschaften in Göttingen* 1 (1771), p.72-83. For more information on the Europe-wide search for extinct volcanoes and columnar basalts, see Davis A. Young, *Mind over magma: the story of igneous petrology* (Princeton, NJ, 2003), p.16-25.
81. Franz Joseph Kinsky, 'Nachricht von einigen Erdbränden im Ellenbogner Kreise in Böhmen', *Abhandlungen einer Privatgesellschaft in Böhmen* 2 (1776), p.58-73. I am indebted to Dr Teadora Shek Brnardić, of the Croatian Institute of History, for this reference.
82. Barthélemy Faujas de Saint-Fond, *Recherches sur les volcans éteints du Vivarais et du Velay* (Grenoble, Joseph Cuchet, 1778); Jean-Louis Giraud-Soulavie, *Histoire naturelle de la France méridionale*, 7 vols (Paris, Jean-François Quillot *et al.*, 1780-1784). On Soulavie's novel ideas on the geological ages of the Earth, see Martin J. S. Rudwick, *Bursting the limits of time: the reconstruction of geohistory in the age of revolution* (Chicago, IL, 2005), p.215-27.
83. Thomas Pennant, *A Tour in Scotland, and voyage to the Hebrides 1772* (Chester, John Monk, 1774), p.299-309. See also chapter 6 in the present volume. For a discussion of the importance of the influential visual exposition of prismatic basalts in the works cited above by Desmarest, Faujas de Saint-Fond and Pennant, see David McCallam, 'From analogies to patterns: images in French and British geological texts (1760-1800)', in *Visualizing the text: from manuscript culture to the age of caricature*, ed. Lauren Beck and Christina Ionescu (Newark, DE, 2017), p.203-27.
84. Barthélemy Faujas de Saint-Fond, *Voyage en Angleterre, en Ecosse et aux îles Hébrides*, 2 vols (Paris, H. J. Jansen, 1797), vol.2, p.34-73 (Staffa), 291 (Edinburgh castle).

wherever one looked in Europe traces of ancient volcanism were being identified and inventoried. Old lava flows, cones of extinct volcanoes and regular basalt columns were all sought as markers of prehistoric volcanic activity and when found were loudly trumpeted as such.

For the amateur geologists of the age, argues Martin Rudwick, scouring these volcanic sites in person was akin to a geological rite of passage. Fieldwork became the secular equivalent of a pilgrimage, at once founding a community of like-minded souls (as among pilgrims) and fulfilling personal quests (as among pilgrims).[85] The result of this collective search for volcanic remains was the slow but sure elaboration of protocols of scientific observation. An empirical gaze was constructed, for which the model may well be that of Sir William Hamilton, posted at the foot of Vesuvius or scrambling among its fiery lava flows. A dare-devil climber, but a careful observer, Hamilton proceeded by observation, comparison, tentative deduction and generalisation only of the most clearly recurrent phenomena. His own credo, established in 1767, was to limit himself to describing events occurring 'under my own inspection, [leaving] their explanation to the more learned in Natural Philosophy'.[86] Twelve years later, he was still insistent on confining himself to 'simple matter of fact, in plain prose, and endeavor to convey to you [...] as clearly, and distinctly, as I am able, what I saw myself, and the impression it made upon me at the time, without aiming in the least at a flowery stile'.[87] When he found himself unable to attend to the volcano's activities, Hamilton paid £20 a year to a local monk, Father Antonio Piaggi, to keep a diary of Vesuvius' daily motions; this the monk did dutifully for nearly fifteen years from his small house at Resina above the excavations at Herculaneum. Hamilton's guide on Etna during his visit there in 1769, canon Giuseppe Recupero, similarly provided him with regular eyewitness accounts of the Sicilian volcano's evolutions.[88] Another correspondent of Hamilton's and a fellow explorer of the volcanoes of

85. Martin J. S. Rudwick, 'Geological travel and theoretical innovation: the role of "liminal" experience', *Social studies in science* 26:1 (1996), p.143-59.

86. Hamilton, *Observations on Mount Vesuvius*, p.1-2.

87. Hamilton, *Supplement to the 'Campi Phlegraei'*, p.5.

88. See Constantine, *Fields of fire*, p.196, 266 (Piaggi) and 55-56, 76-77 (Recupero). Constantine mistakenly gives Giuseppe Recupero, the canon and guide to Etna, the first name of Gasparo. This is, in fact, his brother, a Sicilian jurisconsult. See Andrea Candela, 'Recupero, Giuseppe', in *Dizionario biografico degli Italiani*, vol.86 (2016), http://www.treccani.it/enciclopedia/giuseppe-recupero_(Dizionario-Biografico)/ (last accessed 7 December 2018).

southern Italy, Déodat de Dolomieu, offers a further example of this methodical, empirical approach. In works on the volcanic tephra of the Aeolian Islands and on the cataclysmic Calabrian earthquake of 1783, he bases his analyses on research done *in situ*, with no predetermined hypothesis in mind but a geological hammer in his hand. Denouncing the remote 'semi-scholarly' writings on the Calabrian earthquake, he also treats with extreme caution the accounts of terrified, panicking locals and hardens his heart to terrible scenes of human affliction in order to devote himself 'solely to the physical effects' of the tremors. If he is ultimately to advance the theory of subterranean passageways linking Etna to the devastated Calabrian terrain, it is only after due consideration of all the local factors and the empirical evidence while also acknowledging the lack of 'direct proofs of his theory'.[89] Towards the end of the century, the empirical, inductive methodologies of the likes of Hamilton and Dolomieu were also adopted by certain grand tourists in their travelogues as a means of validating their own experiences on the volcano. Thus Henry Swinburne in his *Travels in the Two Sicilies* (1790) huffs: 'Whoever has not had the advantage of inspecting an active volcano, should not presume to write upon that subject, as he must unavoidably fall into error.'[90]

Volcanological networks and rival schools of thought

By the 1780s Sir William Hamilton had effectively usurped Buffon's mantle as the 'modern Pliny', at least as far as the study of volcanoes was concerned. In fact, his young bride Emma, later mistress to Lord Nelson, was not averse to teasing him as old 'Uncle Pliny'.[91] Not only were Hamilton's own writings on Vesuvius and Etna invaluable testimonies to their activities and effects, but personally he acted as the effortlessly civil intermediary for a host of other volcanological exchanges. His magnificent Palazzo Sessa in Naples was a social, political and scientific hub as were his villas in Caserta, to the north

89. Déodat de Dolomieu, *Mémoire sur les tremblemens de terre de la Calabre pendant l'année 1783* (Rome, Antoine Fulgoni, 1784), p.15, 42, 34, 69. Cited in Simone Messina, 'Le naturaliste et la catastrophe: Dolomieu en Calabre, 1784', in *L'Invention de la catastrophe au XVIIIe siècle: du châtiment divin au désastre naturel*, ed. Anne-Marie Mercier-Faivre and Chantal Thomas (Geneva, 2008), p.285-302. On conceptions of the Calabrian earthquake more generally, see Augusto Placanica, *Il Filosofo e la catastrofe: un terremoto del Settecento* (Turin, 1985).
90. Swinburne, *Travels in the Two Sicilies*, vol.1, p.85.
91. Constantine, *Fields of fire*, p.29.

of the city, and in Portici, close to the royal palace and less than two miles from the foot of the volcano. Familiar with baron Riedesel, the explorer of Etna in 1767, Hamilton emulated the baron's ascent of this volcano two years later and his account of the climb was appended to Riedesel's tour of Sicily when this was published in French in 1773. When Horace-Bénédict de Saussure climbed the same peak also in 1773, the Swiss mountaineer naturally visited Hamilton on his return to Naples and scoured the slopes of Vesuvius with him, just as Hamilton, returning from England a year earlier, had seen nothing untoward about calling on Voltaire at Ferney, later having his study of volcanoes delivered to the great man and corresponding with him on the subject. Hamilton similarly entered into correspondence with his counterpart John Strange in Venice on the volcanic terrain discovered there, with John Hall in Edinburgh and Joseph Banks in London. And if Hamilton regularly accompanied his eyewitness reports to Banks, as president of the Royal Society, with packages of lava samples and volcanic debris, he had himself been shown examples of spun-glass lava by Buffon while passing through Paris. Lava was similarly at the heart of his correspondence with Rudolf Raspe in 1769 when the German asked him to verify Desmarest's theory of the volcanic origin of basalt.[92]

Hamilton also provided the alleged inspiration behind the volcanological expeditions of Patrick Brydone to Etna and Déodat de Dolomieu to the Aeolian Islands. Dolomieu, in turn, established personal relations with Italian volcanologists such as Giuseppe Gioeni, Giovanni Arduino and Alberto Fortis while his *Catalogue raisonné des produits de l'Etna* (1788) was cited frequently in Lazzaro Spallanzani's important *Viaggi alle Due Sicilie*, published between 1792 and 1797, recounting its author's 'volcanological pilgrimages' among the volcanoes of southern Italy.[93] For three summers and autumns in 1788, 1789 and 1790, this professor of natural history at the University of

92. On Hamilton's web of volcanological connections, see Constantine, *Fields of fire*. For the Buffon reference, see Hamilton, *Supplement to the 'Campi Phlegraei'*, footnote on p.7; for Raspe's correspondence, see Young, *Mind over magma*, p.22.

93. Lazzaro Spallanzani, *Viaggi alle Due Sicilie e in alcune parti dell'Appennini*, 6 vols (Pavia, Baldassare Comini, 1792-1797); on his 'pellegrinazioni', see vol.1, p.xxxi. On Spallanzani's geology more generally, see Ezio Vaccari, 'Spallanzani e le scienze geologiche del Settecento: un percorso interpretativo tra carteggi e diari di viaggio', in *La Sfida della modernità: atti del convegno di studi nel bicentenario della morte di Lazzaro Spallanzani*, ed. Walter Bernardi and Marta Stefani (Florence, 2000), p.293-317.

Pavia scoured Vesuvius, Etna, Stromboli, Vulcano and the Aeolian Islands applying Dolomieu's painstaking empirical methodology to the volcanic terrain. One of his most significant discoveries was that of prismatic basalt columns in the smoking, hissing crater of Vulcano.[94] While these articulated prisms, fused with lava on the inner wall of the volcano, were irrefutable proof of basalt's volcanic origins, Spallanzani was so wary of extrapolating theories from his fieldwork that he conceded that some other northern basalts might yet be non-igneous in nature.[95] On Stromboli he climbed to the gushing vent known as the Sciara del Fuoco on the northern side of the volcano's peak. A famous cut-away image in the second volume of the *Viaggi* captures the volcanologist cleaving to a narrow ledge overlooking the spouting, fiery fountains of lava as testament to his scientific daring.[96] Spallanzani also seems to have subscribed to Dolomieu's idea of the subterranean and submarine interconnectedness of the volcanoes in southern Italy. Thus he analysed not just their volcanic products but also the geographical distribution and orientation of their arcs and archipelagos, discerning in their alignment something akin to Buache's mountain 'chains', and similarly conceiving of them as a geophysical 'ossature'.[97] Nonetheless, he was prepared to challenge Dolomieu's thinking when necessary. Hence, he rejected the Frenchman's ideas on volcanic fire, and persisted in carrying out innumerable laboratory experiments on volcanic rock samples to understand better the processes of fusion. His melting of certain lava samples with a powerful reverberatory furnace, used for smelting metals, thus proved the high temperatures generated by volcanism.[98]

Other volcanological networks cut across those of more established sciences, such as astronomy. In this way, Hamilton's predecessor as resident expert on Vesuvius, the Italian priest Giovanni Maria Della Torre, guided the French astronomer Joseph Jérôme Lalande among the lava flows and scoria in 1765.[99] Lalande also frequented Hamilton during his stay in Naples and climbed the volcano with the English ambassador.[100] As an important figure in astronomical

94. Spallanzani, *Viaggi alle Due Sicilie*, vol.2, p.189-92.
95. On the latter point, see Young, *Mind over magma*, p.39, 44.
96. Spallanzani, *Viaggi alle Due Sicilie*, vol.2, p.60-65, plate 3.
97. Spallanzani, *Viaggi alle Due Sicilie*, vol.1, p.xiv.
98. On Spallanzani and his experiments in melting lava samples, see Young, *Mind over magma*, p.265, and Pyle, *Volcanoes: encounters through the ages*, p.95.
99. Lalande, *Voyage d'un François*, vol.7, p.167-78.
100. Constantine, *Fields of fire*, p.29.

circles, the Frenchman was undoubtedly apprised of the astronomer royal William Herschel's sensational discovery in April 1787 of what appeared to be three active volcanoes on the moon. Herschel's lunar volcanoes were said at the time by Hamilton to have put 'Nous autres Volcanistes' firmly in the shade.[101] Herschel, in turn, had further connections with the European geological fraternity, since he frequently conversed with the Swiss geologist Jean-André de Luc when both were living in Windsor in the 1780s in the pay of the royal household. A correspondent of fellow Swiss luminaries such as Jean-Jacques Rousseau and Horace-Bénédict de Saussure, de Luc also chose to publish his work on geological chronology in the form of polemical open letters to James Hutton in Scotland and Johann Blumenbach in Göttingen.[102]

All of which goes to demonstrate that an international cast of established earth scientists, institutional correspondents and amateur volcanologists were in communication with one another, in person and on paper, in the latter part of the eighteenth century. What is more, they all found an increasingly receptive forum in the widely circulating scientific reviews of the day. In addition to the *Mémoires* and *Philosophical transactions* of the state-sponsored academies in Paris and London as well as the local organs of the proliferating provincial bodies imitating them, respected journals such as the *Journal des sçavans* or the *Observations sur la physique, etc.* carried an ever greater number of articles on volcanism or accounts of volcanic exploration.[103] An age of cosmopolitan geology had evidently dawned, but it would be wrong to assume that scientific views on volcanoes were consequently to be brought smoothly into harmony – far from it.

By the time de Luc first hesitantly used the term 'geology' in his *Lettres physiques et morales, sur les montagnes* in 1778, the water-dominated explanations of volcanic activity, reiterated and expanded by Buffon that very year in his *Epoques*, were being seriously called

101. William Herschel, 'An account of three volcanos in the moon', *Philosophical transactions* 77 (1787), p.229-32; Herschel's discovery is cited in Lalande's *Abrégé d'astronomie*, with the claim that one of the lunar volcanoes was seen to flare into life again in 1794 (*Abrégé d'astronomie*, 2nd edn, Paris, Firmin Didot, 1795, p.358). Hamilton, in a letter to Sir Joseph Banks, in Constantine, *Fields of fire*, p.160.

102. See Martin J. S. Rudwick, *The New science of geology: studies in the earth sciences in the age of revolution* (Aldershot, 2004), ch.6: 'Jean-André de Luc and nature's chronology', p.1-19.

103. Marcil, 'Découvrir, comprendre, ressentir la montagne', p.147-49.

into question.[104] As we have seen, it was specifically the volcanological fieldwork undertaken by the likes of Faujas de Saint-Fond, Dolomieu, Hamilton and Raspe that undermined the credibility of Buffon's theorising. Through the 1780s a fundamental division thus began to emerge between two conceptual approaches to earth science, and to the importance of volcanism within this nascent discipline. Two distinct theoretical camps crystallised around the major agency of geological change that each privileged. They also acquired esoteric labels in the process. So, on the one side there were the Neptunists, the school of de Maillet, Buffon, de Luc and Guettard (until 1779 at least) who championed the ubiquitous force of water in shaping the globe. From about 1780 onwards the most influential spokesman of this school of thought was the German mining and mineralogy professor Abraham Gottlob Werner. On the other side stood the Plutonists (also called Vulcanists) who believed in the primacy and creative power of fire. They were headed from 1785 onwards by the Scottish landowner James Hutton. Crudely put, Neptunist geology is chemical, evolutionary and unidirectional – the shrinking of the primitive universal ocean and the cooling of the Earth's core being one-way processes. Plutonists, by contrast, were more mechanistic, revolutionary and cyclical in their thinking. For instance, in his seminal *Theory of the Earth* (1795) Hutton argues that the molten matter of the Earth's core dilates and rises in pockets, forcing a way to the surface to form mountains and specifically volcanoes. Hence the face of the Earth is primarily fashioned by the evacuation of these dangerous subterranean energies in a continually renewed cycle of pressure and release. Geophysics, not chemistry, then determines the principal physical features of the planet.[105] The increasingly bitter debate between the camps led by Werner and Hutton was to drag on well into the nineteenth century before it was more or less settled in favour of the Plutonists.[106]

As for volcanism, Neptunists maintained that seawater was often required to spark it, and that, in the overall chronology of the

104. Jean-André de Luc, *Lettres physiques et morales, sur les montagnes et sur l'histoire de la Terre et de l'homme* (The Hague, Detune, 1778), p.viii.

105. James Hutton, *Theory of the Earth, with proofs and illustrations*, 2 vols (Edinburgh, William Creech; London, Messrs Cadell, Junior, and Davies, 1795).

106. On the Neptunist/Plutonist debate, see Oldroyd, *Thinking about the Earth*, p.97-102; Young, *Mind over magma*, p.16-61; and Mott T. Greene, *Geology in the nineteenth century: changing views of a changing world* (Ithaca, NY, and London, 1982), p.19-45.

Earth's evolution, volcanic phenomena were largely incidental events incapable of effecting the sort of long-term change realised by water erosion and sedimentation. The Plutonists fundamentally disagreed. For them, volcanism was the primary agency shaping the globe. And one by one they countered the Neptunist arguments:

i. the sea is far from the celebrated volcanoes of the Andes, with no possible underground communication with them which might trigger eruptions – hence seawater cannot be the principal catalyst of volcanism;
ii. pyrite is not consistently present in all lavas, while pitch and coal seams are found in large and unstable quantities in areas utterly free of volcanic activity – hence pyrite combustion of carbon-rich deposits is not a viable explanation for volcanism;
iii. vast areas of the European continent are formed of rocks such as basalt which have been proven to be of volcanic origin – hence erosion and sedimentation are at best complementary forces in rock and landscape formation.

The French travel writer Pierre Jean-Baptiste Legrand d'Aussy, reflecting on the basaltic relief of Auvergne in 1794, recapitulated many of these Plutonist arguments, but ultimately he was not convinced that contemporary geological speculations of either camp were anything more than that – mere speculations. 'Too many conjectures are currently bandied about' concerning volcanoes, he writes, that the fables of the Ancients often appear more honest and credible in comparison.[107] And it was certainly true that the neat division of Neptunist from Plutonist proved in many instances to be oversimplified and misleading. For example, Buffon's 'actualism' – his insistence that the same geological processes, such as volcanism, were operative and observable today precisely as they had been throughout time – was a mainstay of Hutton's Plutonist *Theory of the Earth*. In fact, sharing Buffon's antipathy towards orthodox biblical chronology, Hutton famously maintained that the evolutions of the planet's surface had 'no vestige of a beginning, – no prospect of an end'.[108] The 'actual' was also seemingly the 'eternal' natural order. Nor was it a preference for the hard evidence of fieldwork over the easy conjectures of theory

107. Pierre Jean-Baptiste Legrand d'Aussy, *Voyage fait en 1787 et 1788, dans la ci-devant Haute et Basse Auvergne*, 3 vols (Paris, Imprimerie des sciences et arts, 1794), vol.3, p.26-52 (41-42).
108. Hutton, *Theory of the Earth*, vol.1, p.200.

which separated Plutonist from Neptunist. Werner, for example, based his fledgling earth science of 'geognosy' on his own experience of mining in and around his university town of Freiburg. The clearly stratified sediments through which the mines were drilled offered Werner a global model of rocks arranged in a succession of five great layers. The lowest two layers and some of the third were successively deposited, he claimed, by precipitation from the waters of the universal ocean. Perversely, against mounting evidence to the contrary, Werner insisted that granite and basalt were still to be counted among these sedimentary rock types. Volcanism for him arose from the combustion of coal seams, such as those he regularly visited near Freiburg, when water percolated into adjoining pyrite deposits, fermenting and igniting them.[109] The theory might still be an amalgam of Buffon and Lémery, but the fieldwork was Werner's own, and it neatly illustrated *in situ* his popular teachings. Saussure was similarly guided by his great familiarity with the non-volcanic Alps to hold to Neptunist ideas, despite having toured Vesuvius with Hamilton, scaled Etna and studied its lava flows. Thus marine fossils found in abundance on Alpine peaks, such as the immaculately preserved bivalves found on Mont Salève, inclined Saussure to privilege the world-moulding powers of a universal sea over those of underground fires.[110] Even a student of Auvergne, with local knowledge and a hands-on appreciation of its smooth, heat-fused rocks, such as François-Dominique de Montlosier, could assert that fire was not the primary agent of volcanism, despite the evidence around him. Instead, claiming that the nature of volcanic heat eluded contemporary science, he fell back on electrical theories of 'underground thunder and lightning' as furnishing the best available explanation for eruptions and their outpourings.[111] In short, fieldwork could be just as deceptive and unavailing as theory in conceptualising volcanoes.

If intimate knowledge of Alps and mines tended to persuade the likes of Saussure, de Luc, Rousseau, Werner and Lalande of the

109. Abraham Gottlob Werner, *Kurze Klassifikation und Beschreibung der verschiedenen Gebirgsarten* (Dresden, Walther, 1787).

110. See especially Horace-Bénédict de Saussure, *Voyages dans les Alpes*, 4 vols (Neuchâtel, Samuel Fauche, 1779-1796), vol.1, p.190-94; see also Jean-Pascal Alcantara, 'De neige et de lave mêlées: le paysage volcanique d'après Horace-Bénédict de Saussure', in *L'Invention du paysage volcanique*, ed. Dominique Bertrand (Clermont-Ferrand, 2004), p.83-102.

111. François-Dominique de Reynaud de Montlosier, *Essai sur la théorie des volcans d'Auvergne*, 2nd edn (Paris and Clermont-Ferrand, 1802), p.166-68.

validity of Neptunist arguments, the acceptance of some Plutonist precepts was nonetheless not to be ruled out of hand. Thus Saussure, having explored the extinct volcanoes of Auvergne, concurred with Desmarest's idea of the volcanic origin of basalt. Elsewhere in his work, he tried to move beyond the conflicting Neptunist and Plutonist theories of mountain formation by advancing, but never finalising, the notion of the folding back of opposing strata to create mountain ranges, a notion which ultimately proved more accommodating to Neptunist doctrine (rippling of the layers of sedimentation) than to Plutonist thought (subterranean swellings of the crust).[112] Among those more familiar with volcanic landscapes, there was likewise no automatic adherence to Plutonist theory. Hamilton and Dolomieu remained waverers between the two schools of thought until well into the 1780s. In his geological contributions to Jean-Claude Richard de Saint-Non's *Voyage pittoresque, ou Description des royaumes de Naples et de Sicile*, Dolomieu toed a clearly Neptunist line on volcanism drawn largely from Buffon, concluding tentatively with an 'analogy' between volcanic eruptions and electrical storms, but leaving open the question of cause and effect.[113] This work dates largely from 1781. By 1788, however, Dolomieu's study of the volcanic 'pumice islands' (*îles ponces*) off the coast of southern Italy and his account of the 1787 eruption of Etna both contain an explicit rejection of Buffon's Neptunism and a provocative call to revise the latter's *Epoques* to place volcanism in the first period of Earth history as an original and fundamental geological process. Foregrounding the action of 'subterranean fires' in these texts, Dolomieu was not as yet, however, a confirmed convert to Plutonism, and even dismissed the 'monastic squabbling' between the rival schools of geological thought as fatuous and counterproductive.[114] Yet as the century drew to a close, exploring Auvergne lava flows in 1798, Dolomieu finally rallied unequivocally to the Plutonist cause.[115] By that time fellow Plutonists like Raspe, Faujas de Saint-Fond and Breislak, as well as the British contingent of Hutton, Hall and John

112. See Albert V. Carozzi, 'Horace-Bénédict de Saussure pionnier en 1784 du concept de "refoulements horizontaux en sens contraires" en géologie structurale', *Travaux du Comité français d'histoire de la géologie* 3:11 (1977), p.89-112.

113. Saint-Non, *Voyage pittoresque*, vol.2, p.150-56.

114. Déodat de Dolomieu, *Mémoire sur les îles ponces*, p.24-26.

115. See Déodat de Dolomieu, 'Rapport fait à l'Institut national, par le C. Dolomieu, ingénieur des mines, sur ses voyages de l'An v et de l'An vi', *Journal des mines* 46-47 (1798), p.385-402, 405-32. Also the interesting note on this report in Rudwick, *Bursting the limits of time*, p.367.

Playfair, had done much to shake the theoretical foundations of Werner's Neptunism. As a final irony, it was as much Werner's own students as it was foreign opponents of his theory who helped to overthrow the Neptunist intellectual edifice. Chief among these were Leopold von Buch and Alexander von Humboldt whose respective prospecting among the volcanoes of southern Italy, as later among those of the Canary Islands and the Andes, convinced them of the fallacious, parochially determined nature of Werner's Neptunism.[116]

Yet as with all the scientists, scholars and travellers considered in this chapter, the gaze of Neptunist and Plutonist alike, when contemplating the volcano, was far from being unalloyed, objective or rationally disinterested. It was inevitably informed by the powerful aesthetic, philosophical and political discourses of the age. As subsequent chapters will show, these combined to allow the volcano to shape the world anew in ways well beyond the strictly geophysical. The early earth scientists, like their contemporary philosophers, political thinkers and theorists in aesthetics, found in the primordial and eternal volcano an inspiration from more modern thought.

116. See Young, *Mind over magma*, p.51-53.

3. On the volcanic sublime, its art and artifice

Conceptually the notion of the sublime derives from classical studies in rhetoric. The founding text in this tradition, Longinus' first-century-CE work *Peri hypsous*, or *On the sublime*, defines its subject as a form of 'grandeur' in discourse, one which: 'produces ecstasy rather than persuasion in the hearer [...] This is because persuasion is on the whole something we can control, whereas amazement and wonder exert invincible power and force and get the better of every hearer.'[1] Longinus' text was first published in the early modern period by the Florentine humanist Francesco Robortello in 1554. However, its greatest and most influential classical expression was to be found in Nicolas Boileau's French translation, *Traité du sublime*, of 1674.[2] Boileau identifies the sublime as an indefinable power of language, an instance of *je ne sais quoi*, 'something extraordinary and something marvellous which strikes through discourse, and which allows a work to carry away, ravish, transport' its listener or reader.[3] Quite simply, the sublime *moves* its audience; the irresistible force of its rhetoric takes us outside or beyond ourselves. This is the etymological sense of Longinus' 'ecstasy', the *ekstasis* generated by the sublime.

1. Longinus, *On sublimity*, translated by Donald A. Russell (Oxford, 1965), p.2. For more information on the philosophy and aesthetics of the sublime, among a large literature devoted to the subject, see Baldine Saint Girons, *Fiat lux: une philosophie du sublime* (Paris, 1993), and Robert Doran, *The Theory of the sublime from Longinus to Kant* (Cambridge, 2015).
2. For the European reception of the sublime before Boileau, see *Histoire de la rhétorique dans l'Europe moderne, 1450-1950*, ed. Marc Fumaroli (Paris, 1999), p.454-63; and Emma Gilby, *Sublime worlds: early modern French literature* (London, 2006).
3. Nicolas Boileau-Despréaux, *Traité du sublime*, in *Œuvres complètes* (Paris, 1966), p.331-440 (338).

Through Boileau's influence as the supreme French legislator in matters of classical taste, the early eighteenth century inherited a conception of the sublime which was firmly rooted in the field of rhetoric. Yet in Longinus' original, in Boileau's translation, and in the proliferation of other translations of Longinus in the eighteenth century, the sublime was often conceived of as a force of nature, albeit a figurative one.[4] As Chloe Chard has pointed out, the Longinian sublime favoured figures of rhetorical excess, such as hyperbole, which signifies a literal 'throwing beyond' or 'casting out or over'. This, in turn, led it to illustrate its arguments with examples drawn from those mighty natural phenomena which 'actively transgress their own boundaries', such as the Nile in flood or Etna erupting.[5] William Smith's frequently reprinted English translation of *On the sublime* (1739) compares the impassioned oratory of the Athenian Demosthenes with the Roman Cicero in terms of contrasting forms of fury and flames. Demosthenes' vehemence 'seems to be all Fire' and 'may justly be resembled to a Thunderbolt or a Hurricane', whereas Cicero's speech 'like a wide Conflagration, devours and spreads on all Sides'. Smith follows Longinus in then developing the distinction between our everyday indifference towards a 'small Fire' in a 'private Hearth' and our unstinting amazement at the 'celestial Fires' above us; a distinction that is amplified when the same small household fire is set against our awestruck wonder at 'the boiling Furnaces of *Ætna*, which cast up Stones and sometimes whole Rocks from their labouring Abyss, and pour out whole Rivers of liquid and unmingled Flame'.[6] Already in Longinus and his eighteenth-century translators, the element of fire facilitates a subtle shift from a rhetorical to a natural sublime, culminating in the figure of the erupting volcano.

Other eighteenth-century evocations of the sublime directly establish natural phenomena as autonomous manifestations of sublimity, as in François Cartaud de La Villate's *Essai historique et philosophique sur le goust* (1736). Cartaud explicitly inverses the poetic and natural forms of sublime, with the loftiest ideals and images of the Ancients drawn

4. Doran, *The Theory of the sublime*, p.82-94, discussing Longinus' sole volcanic example of an eruption of Etna, p.85-86.
5. Chloe Chard, *Pleasure and guilt on the Grand Tour: travel writing and imaginative geography 1600-1830* (Manchester, 1999), p.111-12.
6. Longinus, *Dionysius Longinus on the sublime: translated from the Greek, with notes and observations, and some account of the life, writings and character of the author, by William Smith* (London, J. Watts, 1739), p.34, 85.

unfailingly from nature, since 'nature is our universal prototype'. It is thus only nature, he writes, 'which furnishes the great objects which constitute the sublime'. It bursts forth in 'the noise of thunder, the view of a great storm, the radiant dawn, earthquakes, awful abysses, vast phenomena, the stately round of the stars, the majesty of the heavens'.[7] Perhaps not fortuitously, this aesthetic shift is mirrored in the evolution in meaning of the Greek term 'catastrophe' in the same period. Originally signifying a tragic reversal of fortunes in the final act of a play, the sense of 'catastrophe' slipped from the theatrical lexicon into that of 'real life' beyond the stage, coming to define great personal misfortunes and major natural disasters – as it still does today.[8] It too, in other words, passed from art to nature in the course of the eighteenth century, albeit that this 'nature' itself remains a fundamentally aesthetic construction.

In its own etymology from the Latin *sublimis*, the sublime intimated a sense of 'elevation' or of 'exalted position' which perhaps inclined it to seek out natural correlatives in mountain landscapes, including those of volcanoes.[9] This inclination was certainly borne out towards the end of the eighteenth century by the relentless extolling of the Scottish Highlands, the Alps and the slopes of Vesuvius and Etna as prime sites of natural sublimity.[10] Nonetheless, this was an association rooted more in an unprecedented growth of travel experiences in these regions than in contemporary theoretical writings on the

7. François Cartaud de La Villate, *Essai historique et philosophique sur le goust* (Amsterdam, n.n., 1736), p.229-30. For more context to Cartaud's ideas on the sublime, see Lawrence Kerslake, 'Aspects of the epistemology of the sublime in seventeenth- and eighteenth-century French theories', *Phrasis: studies in language and literature* 46 (2005), p.7-29.
8. Michael O'Dea, 'Le mot "catastrophe"', in *L'Invention de la catastrophe au XVIII*[e] *siècle*, ed. A.-M. Mercier-Faivre and C. Thomas, p.35-48.
9. Doran, *The Theory of the sublime*, p.39-40. The etymology also draws on *sub* (under, up to) and *limen* (limit, threshold, lintel), a sense of wavering at the limits or occupying an irresolvably liminal position which could apply to the crater's vertiginous rim, neither inside the volcano nor fully outside it. This equates to a more postmodern sense of the sublime, explored in Philip Shaw, *The Sublime* (Abingdon, 2006), p.1, 115-20.
10. The classic study here, first published in 1959, is Marjorie Hope Nicolson, *Mountain gloom and mountain glory: the development of the aesthetics of the infinite* (Seattle, WA, 1997). More recent critical works, with chapters devoted to volcanic mountain terrain, include Remo Bodei, *Paesaggi sublimi: gli uomini davanti alla natura selvaggia* (Milan, 2008), and Cian Duffy, *The Landscapes of the sublime 1700-1830: classic ground* (Basingstoke, 2013).

sublime. For the two most impressive and influential theories of the sublime in the period have almost nothing to say of its mountainous manifestations; what they do instead is to dictate the character of the experience itself. These theories are what we might crudely call the political sublime of Edmund Burke and the moral sublime of Immanuel Kant.

Eighteenth-century theories of the sublime: Burke and Kant

Edmund Burke published *A Philosophical enquiry into the origin of our ideas of the sublime and the beautiful* in 1757 while still in his twenties. A second edition appeared two years later. The *Enquiry* marks a watershed in conceptions of the sublime insofar as it presents it not as a product of rhetorical exaltation or vehemence but as a physiological and psychological transformation of terror into delight. The sublime for Burke is the fruit of feelings not words. It stems from confronting situations 'fitted in any sort to excite ideas of pain, and danger', that is, those 'productive of the strongest emotion which the mind is capable of feeling'. By consequently surviving such situations, he claims, one derives a profound 'negative' pleasure from the absence or removal of apparent threats of intense pain or immediate danger.[11] The sublime is then a psychosomatic sublimation of raw terror into refined delight. Key to this distilling of real pleasure from imagined pain is distance from its object, being at a safe remove from the source of terror. Burke's *Enquiry* goes on to illustrate his theory with social, political, literary and natural experiences of the sublime, as he defines it. On a cursory reading, it would seem that the natural sublime is the most acute of these experiences, inspiring 'Astonishment' in its subject, that is, the 'state of the soul, in which all its motions are suspended, with some degree of horror', realising thus 'the effect of the sublime in the highest degree'.[12] Yet Burke only makes the most perfunctory of references to those natural phenomena which excite this extreme sensation – earthquakes, the expanse of ocean, lightning, 'vast cataracts, raging storms, thunder' and the starry skies.[13] No further comment is offered on their specific sublimity, and volcanic eruptions are conspicuously absent from his list. It is even more ironic then that his one allusion

11. Edmund Burke, *A Philosophical enquiry into the origin of our ideas of the sublime and the beautiful* (London, 1958), p.39-40.
12. Burke, *A Philosophical enquiry*, p.57.
13. Burke, *A Philosophical enquiry*, p.48, 57-58, 82, 78.

to volcanoes in the *Enquiry* is a belated literary one. Admiring Virgil's skill in creating 'sublime' poetry from words independent of their gross physical objects, Burke declares: 'There is not perhaps in the whole Eneid a more grand and laboured passage, than the description of Vulcan's cavern in Etna, and the works that are carried on there.'[14] Far from being exemplary of the mighty natural sublime, the volcano only inspires the grandest emotions in its poetic, mythological manifestation as the forge of the god of fire. Even more dismissively, Burke confuses the volcano in question: it is beneath Vulcano, not Etna, that Virgil situates Vulcan's forge in his *Aeneid*; Etna is home to the Cyclops' terrible furnaces in his *Georgics*. Denatured and mythologised, Burke's volcano points more towards the louring, smoky visions of the 'satanic mills' of industrial Britain than it does to the imposing natural masses of Etna, Vesuvius or Stromboli.

Instances of the overwhelming natural sublime play only secondary roles in Burke's text because he is first and foremost a social thinker. The natural sublime is therefore eclipsed in importance by another form of the sublime: the political. Sublimity is essentially political because, as Burke states, there exists 'nothing sublime which is not some modification of power'.[15] While the original models of this power may well be natural might or brute force, including that of the volcano, 'the power which arises from institution in kings and commanders has the same connection with terror.'[16] Haunting this passage from the natural to the political sublime, through the social imposition of terror, is the barbarous public torture and execution of Robert-François Damiens in March 1757 for his woefully ineffective attempt on the life of the French king Louis XV. As was graphically recounted in the gazettes of the time, Damiens was taken to the place de Grève, solemnly seared with red-hot pincers, scalded with molten lead and boiling oil, then each limb was attached to a horse whipped to tear him apart, but, when his horribly dislocated body remained stubbornly intact, the executioner had to crudely butcher him before consigning his still-palpitating body parts to a public pyre.[17] These grotesque torments inflicted on 'the late unfortunate

14. Burke, *A Philosophical enquiry*, p.170-71.
15. Burke, *A Philosophical enquiry*, p.64.
16. Burke, *A Philosophical enquiry*, p.67.
17. The most famous account is given in the *Gazette d'Amsterdam* of April 1757 and is quoted at length as the opening to Michel Foucault, *Surveiller et punir* (Paris, 1975), p.9-13.

regicide in France' are tellingly Burke's first example of the terror required to excite the sublime in its beholders.[18] Barbarous or not, man is in a state of society, not of nature, for Burke. Hence even his occasional references to the natural sublime lead only to expositions of their social and political correlatives. For instance, the excessive loudness which 'overpower[s] the soul' and 'fill[s] it with terror' might initially be illustrated with 'the noise of cataracts, raging storms, thunder', but it extends inevitably to the man-made blasts of artillery and the terrible 'shouting of multitudes'. The last example is the most revealing in respect of Burke's political conception of the sublime, for he feels compelled to add here that 'the best established tempers can scarcely forebear being borne down, and joining in the common cry, and common resolution of the croud'.[19] This is the nub of the political sublime for Burke: the crowd that can be sublime to the observer able to stand apart and admire its terrible power can all too readily obliterate this safe distance and swallow up even the most objective spectator in its tumult. This is a dire reversal of Longinus' sublime oratory holding irresistible sway over 'every hearer'. Now the orator's voice is dissolved in the 'common cry' of the clamouring masses. Set out in 1757, Burke's political sublime was a dreadful anticipation of the power of the revolutionary mob that he was to denounce so insistently and bitterly forty-two years later.[20] Ultimately Burke's political sublime is conservative, stoical even. Born of fear, specifically the fear of death, it advocates submission to those greater powers which decide over life and death, since 'pain is always inflicted by a power in some way superior'.[21] Hence its complicity with authority and its preference for inert forms of sublimity – vast expanses, wastes, privations, silence, darkness.[22] This also explains why the explosive, anarchic volcano figures so little in his theory of the sublime.

Burke's *Enquiry* was widely read on the continent and clearly influenced thinkers as diverse as Denis Diderot and Immanuel

18. Burke, *A Philosophical enquiry*, p.39; see the fascinating article by Charles Hinnaut, '"The late unfortunate regicide in France": Burke and the political sublime', *1650-1850: ideas, aesthetics, and inquiries in the early modern era* 2 (1996), p.111-36.
19. Burke, *A Philosophical enquiry*, p.82.
20. See chapters 4 and 5 in the present volume.
21. Burke, *A Philosophical enquiry*, p.65.
22. Burke, *A Philosophical enquiry*, p.71: 'All general privations are great, because they are terrible: Vacuity, Darkness, Solitude and Silence'. See also Wessel Krul, 'Style and ideology in Edmund Burke's concept of the sublime', *Phrasis: studies in language and literature* 46 (2005), p.31-41 (35).

Kant.[23] Kant's first study of the sublime was his *Beobachtungen über das Gefühl des Schönen und Erhabenen* of 1764. It is an uneven, immature text, prone to sweeping statements on what constitutes the beautiful and the sublime as a series of self-evident oppositions. Hence night is sublime, day is beautiful; the 'fair sex' is beautiful, the 'noble sex' sublime, and so on. Kant also stoops to misogyny, chauvinism and racism in his peremptory categorising: women's 'philosophy is not to reason but to feel'; the French and Italian are at best beautiful nations, the English of a noble sublimity, but the Spanish of a terrifying cast of the sublime; while 'blacks' are irremediably stupid and so talkative 'that they must be chased apart from each other with thrashings'.[24] Nor does the natural sublime eclipse these fatuous judgements, but figures only as occasionally impressive sights and sounds capable of arousing 'pleasure but with horror'.[25] Moreover, Kant seems to make little distinction between instances of the sublime in art and nature, listing indifferently 'the sight of a mountain whose snow-covered peak rises about the clouds, the description of a raging storm, or Milton's depiction of the infernal kingdom'.[26] Yet where Kant's understanding of the sublime in the *Beobachtungen* gains in power and acuity is in his conception of it as a uniquely moral quality, in which sensations, even extreme ones, are 'ordered under principles'.[27] Here is the intuition of the sublime as the awakening of our moral self-awareness and the possibility of an inner ethical grandeur that exceeds even the most awe-inspiring of nature's physical manifestations. And it is this keenly original intuition of the sublime that Kant was to develop in his later *Kritik der Urteilskraft* (1790), the third of his great philosophical critiques, and a truly influential text in modern aesthetics.

The *Kritik*'s basic lesson is that the sublime (unlike the beautiful) is not to be found in external objects, however imposing they might be,

23. Denis Diderot, *Salon de 1767*, in *Œuvres complètes*, ed. J. Assézat and M. Tourneux, 20 vols (Paris, 1875-1877), vol.11, p.3-367 (especially on the works of Claude Vernet, p.98-148); see also Gillian B. Pierce, *Scapeland: writing the landscape from Diderot's Salons to the postmodern museum* (Amsterdam, 2012), p.63-110.

24. Immanuel Kant, *Beobachtungen über das Gefühl des Schönen und Erhabenen*, in *Werkausgabe*, 12 vols (Frankfurt am Main, 1977), vol.2, p.821-84 (see p.853, 868-70, 880, 882). For a more thorough examination of Kant's understanding of the sublime in this text and in his later *Kritik der Urteilskraft*, see Doran, *The Theory of the sublime*, p.171-285.

25. Kant, *Beobachtungen*, vol.2, p.826.

26. Kant, *Beobachtungen*, vol.2, p.826.

27. Kant, *Beobachtungen*, vol.2, p.840.

but in ourselves, 'only in our own ideas'.[28] Unlike the earlier *Beobachtungen*, ample space is given over here to the natural sublime, but only insofar as the natural sublime affords us the clearest apprehension that, even as it subdues us physically to its might or magnitude, we remain morally independent of, and therefore superior to, the workings of nature. Confronted with storm-tossed seas or dimly distant heavens, our imagination is overwhelmed by their power or their vastness; but our reason intercedes to remind us that we do not solely belong to the physical world of the sea and sky, but can employ concepts such as Freedom, the Infinite or the Divine in order to apprehend, articulate and surpass them. In eighteenth-century terms, the 'sensible' world at its most sublime allows us a glimpse of the 'supersensible' world of the intellect, the exalted realm of principles and ideas. Strong displeasure at our feebleness and finitude in the face of natural phenomena is thereby transformed into intense pleasure at our moral and intellectual pre-eminence over nature.[29]

It is in this context that volcanoes figure in Kant's enumeration of instances of the natural sublime: 'Bold, overhanging and seemingly threatening rocks, thunderclouds towering up to heaven, driven on by crashing noise and flashes of lightning, volcanoes in all their destructive violence, hurricanes leaving devastation in their wake, the boundless ocean rising in fury, the high waterfall of some great river, and suchlike, make our power of resistance insignificantly small in comparison to their might.'[30] However, Kant continues, 'provided we find ourselves in a position of safety', these overwhelming natural forces do not appear terrifying to us, but sublime. He explains this feeling of sublimity as the result of terrible natural phenomena raising 'the strength of our soul far above the ordinary mean, revealing within us an altogether different power of resistance, which gives us the courage to be able to measure ourselves against the seeming omnipotence of nature'.[31] Thus, as long as we behold the erupting volcano – to cite one of Kant's examples – from a safe remove, as Burke too had insisted, then the very failure of our awestruck imagination to *comprehend* the sheer power of the eruption simultaneously inspires an *apprehension* of yet greater moral

28. Immanuel Kant, *Kritik der Urteilskraft*, in *Werkausgabe*, vol.10, p.69-456 (171). See also p.189.
29. See Christoph Madelein and Jürgen Pieters, 'The sublime re-considered: by way of introduction', *Phrasis: studies in language and literature* 46 (2005), p.1-6.
30. Kant, *Kritik der Urteilskraft*, p.185.
31. Kant, *Kritik der Urteilskraft*, p.185.

and intellectual powers within us, powers of an alternative superior order to that of the volcano's dreadful physical energy.

As was his wont, Kant goes on to classify further our experience of the natural sublime, subdividing it into two distinct categories: the 'mathematical' and the 'dynamic'. In short, the mathematical sublime impresses us with its greatness, massiveness or extension, astonishing its beholder with intimations of infinity. As such, it affects particularly the faculty of cognition or reason. The dynamic sublime, by contrast, overwhelms us with an impression of its terrible power or incommensurable energy. It forms the sublime object of the imagination and, crucially for us, includes volcanoes among its natural manifestations.[32] In this way, Kant moralises the natural sublime (as he tended to do with any experience he personally valued), implying that the intense pleasure felt at our moral superiority over nature's merely physical might is a potentially universal attribute of human existence.[33]

Nonetheless, as we have seen in the *Beobachtungen*, he was not above setting sexist, nationalist or racist restrictions on the capacity to savour these special experiences, and so it is with the natural sublime in his third *Kritik*. For its beholder, insists Kant, requires a certain moral culture to appreciate the sublimity of nature at its mightiest or most vast, without which what 'we call sublime only appears to the uncultured mind as frightening'.[34] It is not that the quaking peasant does not feel the sublime, but that he or she is unable to appreciate the universal moral validity of what he or she is feeling. The natural sublime remains subjectively terrifying for such untutored souls without raising an awareness in its beholders of the transcendental, universal power of their own intellect as a shared moral and rational faculty above the merely natural or empirical. To prove his point, Kant cites the trembling Savoyard cattle-herds who look upon the Alps only with fear, and who, according to Horace-Bénédict de Saussure, call intrepid, enlightened mountaineers 'fools'.[35] In practice, if not in principle, this appears to reinforce certain social and cultural limits placed on the category of person who might partake of the natural sublime on Alpine peaks or at the crater of Vesuvius.

32. Kant, *Kritik der Urteilskraft*, p.169-72, 184-89.
33. Malcolm Budd, 'Delight in the natural world: Kant on the æsthetic appreciation of nature. Part III: the sublime in nature', *British journal of æsthetics* 38:3 (July 1998), p.233-50.
34. Kant, *Kritik der Urteilskraft*, p.190.
35. Kant, *Kritik der Urteilskraft*, p.190.

The Alpine sublime and the volcanic sublime

As far as mountains are concerned, Kant prefers to locate the natural sublime in the Alps rather than in volcanic regions. The sweeping vistas, sheer precipices and cloud-piercing summits, producing intimations of the infinite and incarnating the mathematical sublime, are therefore privileged over the explosive, earth-shattering dynamic sublime of the volcano. As if to confirm this preference, Kant cites Saussure elsewhere in his *Kritik* as an example of how the sublime wistful solitude of the mountains is reflected in certain men.[36] Hence the Alps and the volcanoes of southern Italy offer very different forms of sublimity to those who would explore them.

Saussure's contemporary and friend Marc-Théodore Bourrit erroneously perceived evidence of volcanic activity in the Alps during his tour of the Schwyz canton in 1785.[37] But he was on surer ground, following Rousseau and Saussure, in describing the sublime effects of tranquil Alpine sunsets and the simple, free, virtuous character of the mountain peoples. This Alpine sublime of lofty solitude and ethereal freedom not only soars among the snowy peaks but crucially is also present in the lower reaches of the Alps, in homely chalets and 'smiling' valleys. Rousseau in particular tends to view the sublime mountain heights from lusher lowlands as part of delightful valley panoramas.[38] Unlike the grimly isolated volcanic sublime, the Alpine sublime is thus already shifting towards the picturesque. That is, unlike Vesuvius or Etna, the Alps increasingly close, frame and populate their scenes of sublimity. In doing so, they end up denying their beholder the very sensations of infinity, eternity and dizzying extension which constitute the natural sublime. Put another way, if the essence of the sublime is for the beholder to identify with an abstraction (Freedom, the Infinite, the Divine) and so negate his or her individuality in a thrilling first flush of terror, the Alps afford fewer and fewer abstract landscapes to their visitors.[39] With ever more tourists passing seamlessly between

36. Kant, *Kritik der Urteilskraft*, p.204.

37. Marc-Théodore Bourrit, *Nouvelle description générale et particulière des glacières, vallées de glace et glaciers qui forment la grande chaîne des Alpes de Suisse, d'Italie & de Savoye*, 3 vols (Geneva, Paul Barde, 1785), vol.2, p.135-36.

38. Americo Castro, 'Rousseau and the Alps', in *An Idea of history: selected essays of Americo Castro*, ed. Stephen Gilman and Edmund L. King (Columbus, OH, 1977), p.271-76.

39. See Christine Battersby, 'Terror, terrorism and the sublime: rethinking the sublime after 1789 and 2001', *Postcolonial studies* 6:1 (2003), p.67-89 (67-68); see

the florid meadows, placid lakes and barren rocky heights, the Alpine experience in the late eighteenth century becomes the epitome of the picturesque, at least as William Gilpin conceived of it, namely, as an intermediary aesthetics between the extremes of Burke's beautiful and sublime.[40]

It is not so with the volcanic sublime. Framed and peopled, especially in the prints of the age, the volcano nonetheless resists being reduced to the merely picturesque because its particular form of sublimity is not inert, vast, steep or extensive. Volcanoes may leave behind lunar desolation and lava-smooth wildernesses, but their sublime quality lies less in their fathomless craters and barren expanses than in their unpredictable eruptive energy. It is as a figure of chaos unleashed that the volcano is sublime. In this sense, unlike the Alpine sublime, a volcano like Etna gives us a glimpse of nature at war with itself, nature seemingly dissolving into fiery disorder, returning to primeval formlessness. Christopher Hitt has even suggested that the late eighteenth century's appalled fascination for 'wild nature' (which would include the volcano, hurricane, earthquake, tidal wave, etc.) represents one of the first faltering expressions of our current ecological consciousness; that the catastrophic natural sublime affords us an early, uncertain intimation of geological and climatic chaos.[41] Hitt's notion of a modern 'ecological sublime' revises the more triumphant aspects of the Kantian natural sublime by recentring its sublimity on nature's radically estranging, humbling, unfathomable forces. It does not deny the sublime its crucial transcendence, but unlike Kant, who places this in reason's soaring dominion over brute nature, the ecological sublime

---

also the same author's *The Sublime, terror and human difference* (London, 2007), especially p.1-45.

40. In contrast to the sublime, Burke defines the beautiful as 'that quality or those qualities in bodies by which they cause love, or some passion similar to it'. Elsewhere in his *Enquiry*, the beautiful is described as evoking tenderness, deriving from the apprehension of small, pleasing objects, of smoothness, delicacy or 'gradual variation'. See Burke, *A Philosophical enquiry*, p.91, 112-16. See William Gilpin, *Observations on the River Wye, and several parts of South Wales, etc., relative chiefly to picturesque beauty, made in the summer of 1770* (London, R. Blamire, 1782); also Uvedale Price, *An Essay on the picturesque, as compared with the sublime and the beautiful* (London, J. Robson, 1794).

41. Christopher Hitt, 'Towards an ecological sublime', *New literary history* 30:3 (summer 1999), p.603-23. For an alternative source of early modern environmental consciousness, see Richard Grove, *Green imperialism: colonial expansion, tropical island Edens and the origins of environmentalism, 1600-1860* (Cambridge, 1996).

locates a transcendent moment in overcoming the nature–culture, object–subject dualisms through our unmediated exposure to manifestations of extreme natural might or magnitude. This conception is particularly pertinent to the volcanic sublime insofar as the ecological sublime no longer relates to the 'mathematical' sublimity of nature's eternal permanence and inviolability (how could it, in an age of nuclear accidents, oil slicks and unregulated mining?). Instead, it is recognised in those expressions of the 'dynamic' sublime which are fundamentally natural (volcanic eruptions, earthquakes, floods) but which are also precipitated increasingly by humankind's ongoing destabilisation of geological, ecological and climate systems. The eighteenth-century volcanic sublime thus adumbrates the ecological sublime of the twenty-first century in its experience of unmediated wild nature as both event and matter, or, more specifically, as both overwhelming event (eruption) and excessive matter (ash clouds, lava). The unpredictable immediacy of volcanic activity confounds, and so transcends, our usual categories of imagination and perception of the natural world – it is experienced as *before* and *beyond* the epistemological certainties of nature–culture and other–self.

The volcano as tableau

Distinct from the beautiful which has a form, and from the picturesque which has a frame, the natural sublime in its most elementary manifestations remains an obscure, intangible, inexpressible excess. It not only spills beyond imagination, it also reaches beyond words. The specific challenge therefore posed by the sublime to late-eighteenth-century writers, artists and thinkers was to represent the unrepresentable, to think the unthinkable and describe the indescribable. An initial strategy in this situation was to concede one's all-too-human limitations. Faced with the overwhelming inchoate violence of, say, Vesuvius erupting or Lisbon levelled to the ground, language, and indeed all art, often chose to profess its inadequacies. Thus on the rumbling, lava-streaked slopes of Vesuvius, Jean-Marie Roland de La Platière confesses to his reader: 'I will never be able to paint for you the dreadful horror of this spectacle.'[42] Even so seasoned a volcano observer as Sir William Hamilton had to concur, writing of the 1779 eruption of the same mountain: 'this was more beautiful and sublime than even the most

42. Roland de La Platière, *Lettres écrites de Suisse*, vol.2, p.234.

lively imagination can paint to itself', forming 'such a scene, as no power of art can express'.[43]

With meaning stretched to the point of meaninglessness in the face of the natural sublime, language is often reduced to its most paradoxical or seemingly confused forms. Here oxymorons proliferate. The natural sublime becomes the site of delicious terrors, dreadful joy, horrible ecstasies. These abrupt juxtapositions of opposing qualities are taken to an extreme in the case of the volcano, for, as Hester Piozzi notes of Vesuvius, the volcano is 'guarded as all its secret caverns are [...] with every contradiction'.[44] The starkly contradictory quality of sublime feelings (terrible delight) is physically manifest in the volcano's form. One scales its heights to peer into the depths of the Earth, its freezing summit is the seat of searing flames, its lava is deadly to the living but regenerative of life, ultimately enriching the terrain it smothers. The volcano is at once snow and fire, sterile and fertile, hot and cold, repulsive and alluring, empty and overflowing, gloomily silent and deafeningly loud, hellish and heavenly. It is a world turned upside down and inside out. It is no surprise then that it appealed to minds, such as Sade's, apt to run to carnivalesque extremes. Yet, despite owning despairingly that the volcano is thus beyond words, sublimely inexpressible and unrepresentable, eighteenth-century explorers of its slopes rarely failed to attempt vivid tableaux of what they saw, heard and felt. Let us take two striking examples of this tendency: Patrick Brydone on Etna in 1770 and Charles Dupaty on Vesuvius in 1785.

For his part, Brydone gives himself an explicitly classical model. His party of nine led by the sure-footed Blasio, or 'the Cyclops', were to emulate the 'emperor Adrian [*sic*]' and more controversially 'the philosopher Plato' in reaching the summit of Etna in time to see the sun rise.[45] This meant a nocturnal ascent. Hence, after a hasty supper just before midnight, the party made slow progress on mules up the forested lower slopes. When they eventually broke beyond the tree line, they sent their mules back down the mountainside and began a halting, aching climb through snowfields covered in crusts of ash, with air temperatures plunging. Finally, after a further hour, they reached the crater rim, 'to see the most wonderful and most sublime sight in nature'.[46] Brydone's very use of the term 'sublime' here necessarily

43. Hamilton, *Supplement to the 'Campi Phlegraei'*, p.8-9.
44. Piozzi, *Observations and reflections*, p.350.
45. Brydone, *A Tour through Sicily*, vol.1, p.183.
46. Brydone, *A Tour through Sicily*, vol.1, p.187.

prompts mention of the inadequacy of language to do justice to what they beheld – 'here description must ever fall short'[47] – just as it later incites him to remark on the oxymoronic character of a site where 'regions of intense cold and of intense heat seem forever to be united in the same point.'[48] What confronts them is described in suitably dislocated syntax:

> This point or pinnacle, raised on the brink of a bottomless gulph, as old as the world, often discharging rivers of fire, and throwing out burning rocks, with a noise that shakes the whole island. Add to this, the unbounded extent of the prospect, comprehending the greatest diversity and the most beautiful scenery in nature; with the rising sun, advancing in the east, to illuminate the wondrous scene.[49]

Climbing the volcano at night was precisely intended to heighten the confusion of the senses and intensify the colour and spectacle of spurting, cascading red and yellow lava against the darkness enshrouding the rest of the view. Torch-lit ascents were a popular feature of visits to Vesuvius, too, in the late eighteenth century. Yet Brydone seeks here on Etna to combine and superimpose a number of intimations of the sublime in keeping with his declaration that there is not 'on the surface of this globe, any one point that unites so many awful and sublime objects'.[50] He effectively sets up a stunning contrast between the tumultuous volcanic sublime glimpsed over the crater edge and the 'absolutely boundless' vistas slowly revealed to his party at dawn, an extensive sublimity lost in distant horizons and discovered also from Alpine peaks (as Brydone later suggests on the descent).[51] Similarly, on his way up the volcano, he reaches for awestruck evocations of the 'majesty and splendour' of the 'immense vault of the heavens' glittering with stars shining more brightly than ever before.[52] These prefigure one of Kant's favourite examples of the 'mathematical' sublime – the infinite, unthinkable expanse of the universe revealed to us at night. In other words, Brydone's method for expressing the inexpressible volcanic sublime, of which 'no imagination has dared to form an idea', is to offset it with other instances of sublimity: the 'light and darkness [...]

47. Brydone, *A Tour through Sicily*, vol.1, p.187.
48. Brydone, *A Tour through Sicily*, vol.1, p.192.
49. Brydone, *A Tour through Sicily*, vol.1, p.188.
50. Brydone, *A Tour through Sicily*, vol.1, p.187.
51. Brydone, *A Tour through Sicily*, vol.1, p.190, 197.
52. Brydone, *A Tour through Sicily*, vol.1, p.184.

still undivided' and 'so boundless a scene' viewed from the summit.[53] The volcano is sublime not only because of its unimaginable energy but because its shattered peak provides a singular vantage point from which to seize, in quasi-divine fashion, the totality of the expansive view above and below. Paradoxically, Brydone makes a frame of infinite extensions, the starlit skies above and the limitless prospects all around. Thus embracing the whole vista before them in a single gaze, Brydone's spectator-readers are posted at the centre of a sublime *tableau* which exists at once because of them and independently of them. This unified, instantaneous tableau, called into being by its beholders yet seemingly oblivious to their presence, is the prime vehicle for the volcanic sublime. As such, it conforms to an aesthetic notion of the tableau, as this was theorised by the likes of Diderot in the 1760s and practised by various painters and dramatists of the age.[54]

Charles Dupaty's ecstatic account of his 1785 exploration of Vesuvius makes this point yet more overtly. Crossing the lava fields on the upper slopes of the mountain, Dupaty stops to contemplate the scene:

> Before me the shadows of the night and the clouds were thickening with the smoke issuing from the volcano and floating around its summit; behind me the sun now sunk beyond the mountains spread its dying rays over the coast of Posillipo, Naples and the sea while over the isle of Capri the moon appeared on the horizon; so in that instant I saw the billows of the sea sparkle at once with the light of the sun, of the moon and of Vesuvius. What a beautiful tableau![55]

Dupaty too is struck by the evidently contradictory character of the volcano, by 'that dreadful, sterile, deserted nature and that smiling, animated, fecund nature' starkly juxtaposed.[56] His tableau of Vesuvius at sunset forms a neat diptych with Brydone's description of Etna at sunrise. Yet Dupaty goes one step further in integrating himself and thus his reader into the vivid tableau he draws. Unlike

53. Brydone, *A Tour through Sicily*, vol.1, p.187, 188, 190. If one thought that Brydone's manifold sublimities on Etna would be difficult to outdo, Thomas Watkins, at the peak of the same volcano in March 1788, adds aerial motion to this sublime vision in the form of an imagined, perilous balloon ride over the churning, spouting crater. See Watkins, *Travels*, vol.2, p.17.
54. See Michael Fried, *Absorption and theatricality: painting and beholder in the age of Diderot* (London, 1980).
55. Dupaty, *Lettres sur l'Italie*, vol.2, p.209.
56. Dupaty, *Lettres sur l'Italie*, vol.2, p.209.

Brydone in his depiction of the ascent of Etna, drafted from a sickbed the day after the climb, Dupaty claims to dash down his awestruck impressions 'on the summit of Vesuvius, by the glow of an eruption, at midnight'.[57] The absorption of the spectator-reader into the spectacle is complete; here the tableau includes its creator in the very act of creating it.

Little wonder then that eighteenth-century artists in watercolours, gouache and oils (who not infrequently included themselves in the dazzling canvases they painted of volcanoes) were irresistibly drawn to the play of fire and night, of light and dark, of shades and hues, of mountain, sea and sky, in order to pit their art against the sublime volcano (not to mention turning a handsome profit in the process).

The sublime volcano in art

It is with the seventeenth-century Italian baroque artist Salvator Rosa that volcanoes begin noticeably to intrude on landscape art. Born in Naples, Rosa was familiar with its rugged volcanic wildernesses. His jagged, barren landscapes mark a stark departure from Claude Lorrain or Nicolas Poussin's luxuriant and soft Arcadian scenes. As the poet James Thomson noted in his *Castle of indolence* (1748), Lorrain's warm canvases contrast with what 'savage *Rosa* dash'd, or learned *Poussin* drew'.[58] Rosa's most startling volcanic painting is his *Death of Empedocles*[59] of the 1660s, a work that finds a literary echo in William Beckford's *Biographical memoirs of extraordinary painters* (1780) in which the fictional sixteenth-century Og of Basan, distraught and haunted by his abandoned lover, scrambles up Etna and is seen by local peasants falling 'from a cliff', that is, tumbling into the volcano's crater.[60] Beckford would be all the more familiar with these volcanic

57. Dupaty, *Lettres sur l'Italie*, vol.2, p.206.
58. James Thomson, *The Castle of indolence: an allegorical poem, written in imitation of Spenser* (London, n.n., 1748), p.14.
59. Salvator Rosa, *The Death of Empedocles* (*c.*1665-1670), private collection. On Rosa, see Helen Langdon, Xavier Salomon and Caterina Volpi, *Salvator Rosa: bandits, wilderness and magic* (London, 2010). For the following discussion on the volcano in European eighteenth-century art I am also indebted to the insights in the beautifully illustrated chapter 'Nocturne: Gemalte Gewalt', in Joachim von der Thüsen, *Schönheit und Schrecken der Vulkane: zur Kulturgeschichte des Vulkanismus* (Darmstadt, 2008), p.65-95.
60. William Beckford, *Biographical memoirs of extraordinary painters* (London, J. Robson, 1780), p.85-89.

landscapes, as he had travelled in southern Italy in 1780 and would return, accompanied by the British artist John Robert Cozens, in 1782. Cozens produced over ninety watercolour landscapes for his wealthy patron in the course of this tour, many featuring Vesuvius sketched from Portici and its environs.[61] In fact, the artist–tourist partnership was a major spur to the production and dissemination of sublime volcanic landscapes, with many smaller commissions emulating those made by Goethe from Johann-Heinrich Tischbein, by Sade from Jean-Baptiste Tierce, by the abbé de Saint-Non from Hubert Robert, by Bergeret de Grancourt from Pierre Jacques Volaire, and by Sir William Hamilton first from Pietro Fabris and later from Saviero della Gatta.

A further significant departure from the timeless bucolic landscapes of seventeenth-century painters such as Poussin and Lorrain was the fact that this eighteenth-century volcanic art was not infrequently dated to particular eruptions. For example, Jakob Philipp Hackert claims that one of his spectacular tableaux of *Der Ausbruch des Vesuv* depicts precisely and exclusively the nocturnal lava burst of 12 January 1774 (Figure 1). Having already troubled the Arcadian harmony of an eternal nature, the volcano thus enters into a fraught relationship with the most prestigious artistic genre of the period, namely, history painting. It could be argued that the great Lisbon earthquake of 1755 had already announced the entry of natural catastrophes into contemporary historical consciousness and thus into history as a narrative and artistic mode of testimony. If so, specifically dated paintings of Vesuvius and Etna further dissolved the boundaries between a lived present and a remote or mythologised past. In the second half of the eighteenth century, both artists and their patrons wanted to mark the living moment of their encounter with the sublime volcano and to recognise themselves in the well-dressed, emboldened spectators who never fail to populate the darkened foregrounds of these tableaux. This tendency also popularised the use of gouache, rather than oil paints, to capture the eruption in its immediacy, since gouache allowed for a quicker execution and fuller canvas coverage and was often used in smaller-scale work that

61. See C. F. Bell and Thomas Girtin, 'The drawings and sketches of John Robert Cozens: a catalogue with a historical introduction', *The Volume of the Walpole Society* 23 (1934-1935), p.i-xi, 1-87 (iii-v, vii, ix-xi, 1-25, 27-81, 83-87); the article includes 'Some additions and notes to the twenty-third volume of the Walpole Society', p.1-12.

Figure 1: Jakob Philipp Hackert, *Der Ausbruch des Vesuv in der Nacht des 12. Januar 1774* (1774). Kassel, Neue Galerie.

facilitated subsequent shipping.[62] In contrast, the large-scale, formally historicised depictions of Vesuvius in the early nineteenth century, such as Pierre Henri de Valenciennes's *Eruption du Vésuve arrivée le 24 août de l'an 79* (1813) or John Martin's *The Destruction of Pompeii and Herculaneum* (1822), stand as belated attempts to subordinate once more the volcano's sublime might to the classical historic moment of its eruption, to turn history itself into a picturesque framing device.[63]

However, especially in the latter part of the eighteenth century, the volcano afforded an array of artists an opportunity to intensify aesthetic and emotional contrasts in an implicitly shared present. Claude-Joseph Vernet's paired paintings of a violent storm at sea and

62. See Emilie Beck Saiello, *Pierre Jacques Volaire 1729-1799, dit le chevalier Volaire* (Paris, 2010), p.190-91.

63. Pierre Henri de Valenciennes, *Eruption du Vésuve arrivée le 24 août de l'an 79 de J.-C. sous le règne de Titus* (1813), Toulouse, Musée des Augustins; John Martin, *The Destruction of Pompeii and Herculaneum* (1822), London, Tate Galleries.

Figure 2: Carlo Bonavia, *Veduta del Vesuvio in eruzione dalla lanterna del molo* (1757). Beaulieu, National Motor Museum.

subsequent sunlit calm had already juxtaposed complementary visions of the sublime and the beautiful in nature.[64] Now the erupting volcano (almost always Vesuvius) concentrated in a *single* canvas the same stunning contrasts.[65] This is exemplified in the Italian artist Carlo Bonavia's 1757 painting *Veduta del Vesuvio in eruzione dalla lanterna del molo*, commissioned by the English tourist Lord Brudenell (Figure 2). Anticipating Dupaty's verbal tableau of 1785, Bonavia aligns three sharply distinct light sources: the glimmering lighthouse on the left, the erupting volcano in left-centre and a blaze of moonlight on the right. It opposes the orange and scarlet rivers of the volcano to the cool blue of the night sky, white disc of the moon and dirty-white

64. On Vernet and the sublime, see Nicolas Gaudreau, 'La gloire du peintre et les errements de l'Académie: des pistes pour l'étude des marines de Claude-Joseph Vernet au salon', *RACAR: revue d'art canadienne / Canadian art review* 27:1-2 (2000), p.74-86.

65. The accumulated contrasts could be taken too far, as in Abraham-Louis-Rodolphe Ducros, *L'Eruption du Vésuve et naufrage*, known as *L'Eruption des cendres* (*c*.1800), Lausanne, Musée cantonal des beaux-arts.

moonlit clouds. This contrast is reinforced by the strong geometrical shapes governing the different features of the painting: the pencil-straight lighthouse defying the squat cone of the volcano, with the ancient crater of Mount Somma repeating darkly the conical outline of Vesuvius aflame, both standing apart from the round moon in an oval of woolly cloud. In other paintings of Vesuvius by Bonavia this geometrical arrangement is altered to accommodate a more substantial diagonal stream of lava on the mountain slopes, pouring from the decentred cone of the exploding summit in the top left of the painting to the glinting planes of the still sea and moonlit horizon at bottom right. Pierre Jacques Volaire employs a similar structure in his famous and much-copied *L'Eruption du Vésuve en 1771* (Figure 3),[66] although here repeated cones and irregular triangles dominate the scene and the moonscape is more ragged. These strong diagonals of red-orange lava create the illusion of temporal movement as the gaze travels the length of them from erupting peak to shimmering sea. This is noticeably less the case with the central vertical fountains of yellow lava characteristic of Joseph Wright of Derby's paintings of Vesuvius.[67]

What sets up a further level of contrasts in these sublime images of eruptions at night is the groups of dwarfed spectators inevitably posted in the foreground. Their principal role, of course, is to connote the spectacular nature of what they behold; as Susan Sontag puts it in the *Volcano lover*, 'a spectacle requires the depiction of a gaze.'[68] But in Bonavia's painting of 1757, as in many others of the time, the spectators also divide into frock-coated tourists striking suitably impressed poses of awe and local peasants and fishermen barely paying attention to the eruption. This socially determined contrast is often inflected philosophically too by the inclusion among the human figures of knots of local priests or monks brandishing ineffectual relics or puny crosses at the flaming mountaintop.[69] Volaire and Wright of Derby heighten yet further the distinction between spectator and dreadful spectacle by often casting the former in black silhouette against the bright reds

66. This eruption took place from 1 May to 11 May 1771 and became the archetype for Volaire's most widely reproduced images of Vesuvius erupting. See Beck Saiello, *Pierre Jacques Volaire*, p.171-74.
67. See Nicolson, *Joseph Wright of Derby*, especially vol.1, p.75-85.
68. Susan Sontag, *The Volcano lover: a romance* (London, 1993), p.57. This is a fictionalised account of Sir William Hamilton's life and particularly his love affair with Emma Hart, later his second wife and Lord Nelson's mistress.
69. See chapter 4 in the present volume for contrasting notions of 'enlightenment' here.

Figure 3: Pierre Jacques Volaire, *L'Eruption du Vésuve en 1771* (1771). Chicago, IL, The Art Institute of Chicago.

of the cascading lava. The spectators might similarly be positioned on dramatic crags or rocky outcrops which in reality either do not exist or would be so close to the lava streams as to ensure that the curious observers were incinerated instantly. This is the case with Volaire's gesticulating company on the reconfigured Atrio del Cavallo in *L'Eruption du Vésuve en 1771*. In fact, as Emilie Beck Saiello notes, these groups of amazed spectators were added last to the painting and, in Volaire's case, were often recast to include recognisable portraits of his patrons. In a sense, this attention to populating his explosive volcanic tableaux diminished their raw sublimity, inclining them more towards controlled, framed picturesque compositions.[70]

It is then a small step from changing the volcanic topography for dramatic effect to displacing the volcano wholesale to provide 'sublime' backdrops to other subjects; a small step from including flattering miniatures of the artist's patrons to seating them casually before a Neapolitan landscape featuring a smoking Vesuvius in the distance. Examples of this tendency include Elisabeth Vigée-Lebrun's portrait of Frederick Augustus Hervey, earl of Bristol, or Wright's own portrait attributed to Richard Hurleston.[71] However, the volcanic sublime could be manipulated in the opposite sense too, heightened or intensified for aesthetic effect. And in certain of his twenty-eight views of Vesuvius Wright goes farther than any of his contemporaries in toying with the volcano's searing sublimity. He might retain the classic contrast between the triple light sources found in Bonavia – those of lighthouse, volcano and moon – as in his *Vesuvius in eruption* (1775-1778). In other views of Vesuvius, the relation between the light sources evolves more subtly, with the moon increasingly faint and tainted by the orangey-yellow glow cast by the erupting volcano. Elsewhere the lighthouse is removed, the moon absent, and the volcano defiantly occupies the centre of the tableau, shooting livid yellow jets of lava and flame high into the night sky, illuminating only the rocky wastes around the summit.[72] These last tableaux realise most fully a conception of an overwhelming, excessive sublime.

Wright appears to use volcanic eruptions as an experimental palette of sulphurous yellows, the most vivid of which was actually

70. Beck Saiello, *Pierre Jacques Volaire*, p.163, 134-39.
71. Louise-Elisabeth Vigée-Lebrun, *Frederick Augustus Hervey, 4th earl of Bristol and bishop of Derry* (1790), Ickworth, Suffolk, National Trust Collections; Richard Hurleston, *Joseph Wright of Derby* (1774-1776), Yale Center for British Art.
72. See Nicolson, *Joseph Wright of Derby*, vol.1, p.76-77.

called Naples Yellow and was made by the prolonged roasting of the potentially poisonous oxides of lead and antimony.[73] As such Wright's volcanic eruptions range from the classical sublime of Vernet and Bonavia to views which are tinged with an artificial, indeed an eerily industrial, light. Nowhere is this clearer than in his *Eruption of Mount Vesuvius with the procession of St. Janarius's head* of 1778 intended to stand as a diptych to the blazing *Firework display at the Castel Sant'Angelo* (1778-1779).[74] This latter Roman pyrotechnic extravaganza was also known as the Girandola, and twice annually it lit up the night skies on holy feast days. Sir William Hamilton had already evoked it in relation to Vesuvius in 1766, writing how impossible it was to describe the beauty of the fizzing, scorching debris thrown out by the volcano, which he called 'these girandoles of red hot stones, far surpassing the most astonishing artificial fire-work'.[75]

Yet in Wright's fiery diptych a different relationship obtains between the firework display and the volcanic eruption, since his canvases effectively place them on an aesthetic par. What Wright calls the 'greatest effect of Nature' no longer outdoes the greatest effect 'of Art'.[76] Both are pushed into the distance so that their bright yellows might resonate more indistinctly and equally, and both are framed – the fireworks from the left, the volcano from the right – by dark overhanging trees. If anything, Wright's fireworks burn with a more intense, searing yellow light. In comparison, the volcano's sublime effulgence is somewhat regulated and domesticated. The various framing devices used by Wright in this pairing of Vesuvius and the Girandola (trees, staggered habitations, orange reflections in water, etc.) are only the formal equivalents of more abstract or improvised frames used by other eighteenth-century visitors to Naples, which succeed similarly in restricting the volcanic sublime to a more picturesque

73. See Andrew Moore and Nigel Larkin, *Art at the rockface: the fascination of stone* (London, 2006), p.14, 32.

74. Joseph Wright of Derby, *An Eruption of Mount Vesuvius with the procession of St. Janarius's head* (1778), Moscow, The Pushkin Museum of Fine Art; *Firework display at the Castel Sant'Angelo in Rome (La Girandola)* (1778-1779), St Petersburg, State Hermitage Museum. See the beautiful reproduction of this diptych of volcanic images in Emilie Beck Saiello and Brice Gruet, 'De la girandole du château Saint-Ange au Vésuve en éruption: voyage d'une métaphore visuelle', in *Le Vésuve en éruption*, ed. E. Beck Saiello and D. Bertrand, p.151-67 (162-63).

75. Hamilton, *Observations on Mount Vesuvius*, p.8. See also Beck Saiello and Gruet, 'De la girandole du château Saint-Ange au Vésuve en éruption', p.151-67.

76. Cited in Nicolson, *Joseph Wright of Derby*, vol.1, p.279.

composition.[77] These include the portraitist Elisabeth Vigée-Lebrun recalling the volcano in her memoirs as 'the sight of this superb tableau and all that framed it', or Hester Piozzi using an apartment window to compose and 'command a complete view of the mountain'.[78]

Artificial volcanoes

In fact, Piozzi was well aware not only that the sublime volcano could be reconfigured and reframed artistically, but also that it constituted an eminently movable feast for the senses. Her *Observations and reflections made in the course of a journey through France, Italy and Germany* record that, in 1785, the inhabitants of Bologna hit upon the odd idea of entertaining the visiting king of Naples by 'making a representation of Mount Vesuvius on the Montagnuola, or place of evening resort', necessarily accompanied by 'fireworks'.[79] They had been preceded in this by the authorities of Rome who saw fit to open and close the Girandola firework display of March 1780 with a simulated eruption of Vesuvius, or so reports the artist Antonio Canova.[80] In England in the 1790s William Beckford had similarly considered erecting an artificial volcano in his gothic parklands at Fonthill Abbey before opting for a brooding crater lake modelled on volcanic Lake Nemi near Rome, while night-time pyrotechnics and smoke were also allegedly deployed from the top of Wilhelmshöhe near Kassel in 1799 to imitate an eruption on the occasion of a royal visit.[81]

Yet all of these mocked-up volcanoes appear to be but pale and feeble imitations in comparison with the most famous and spectacular of all fake Vesuviuses of the eighteenth century: the so-called 'Stein' or Stone Island in the English garden at Wörlitz, near Dessau in eastern Germany. In the eighteenth century Wörlitz was home to Prince Leopold III Friedrich Franz of Anhalt-Dessau, an enlightened German nobleman who had visited the Bay of Naples in 1766 as part of his Grand Tour. So awestruck was he by the experience that twenty years

77. For a comparison with another manipulation of the sublime by Wright, see Jesse Molesworth, 'The cosmic sublime: Wright of Derby's *A Philosopher lecturing on the orrery*', *Lumen* 34 (2015), p.109-21.

78. Vigée-Lebrun, *Souvenirs*, vol.2, p.110; Piozzi, *Observations and reflections*, p.306-307.

79. Piozzi, *Observations and reflections*, p.186.

80. See Nicolson, *Joseph Wright of Derby*, vol.1, p.9.

81. On the Fonthill and Kassel 'volcanoes', see Thüsen, *Schönheit und Schrecken der Vulkane*, p.126-27; see also Christopher Thacker, 'The volcano: culmination of the landscape garden', *Eighteenth-century life* 8 (January 1983), p.74-83.

later he began building a replica of the bay in his parklands, including a seventeen-metre-high brick Vesuvius, covered in local boulders and equipped with three fireplaces in its shallow crater. The crater could also be half-filled with water in order to produce lava effects by pumping streams over copper slag backlit by red lamps. It is thought that fresh pine needles were burned to produce crackling clouds of billowing smoke which rose some thirty feet above the summit. Firecrackers added further thunder and lightning. Wörlitz's 'Vesuvius' stood in the south-east corner of a large ornamental lake and its first 'eruption' took place in the summer of 1794 (ironically at precisely the same time that the real volcano was roaring into fiery activity and engulfing the hapless village of Torre del Greco in torrents of lava forty feet deep).[82] An amazed boating party, including German royalty, witnessed the thrilling explosion at night from the lake waters, set adrift in the dark in order to intensify the illusion of being suddenly transported to the Bay of Naples and confronted with the erupting volcano.[83]

Yet for all this elaborate artifice, Leopold was apparently a sincere student of the contemporary debates about volcanism, such as the basalt controversy of the 1770s and 1780s, going so far as to proclaim his own Plutonist belief in basalt's volcanic origins by studding his ornamental lake with dark prismatic columns of the rock allegedly modelled on those depicted off the coast of Sicily in Jean-Pierre Houël's *Voyage pittoresque* (1782-1787).[84] Unfortunately, his ersatz Vesuvius did little to illustrate the primacy of volcanism as an earth-shaping force, since it acted first and foremost as a superficial, if impressive, social spectacle, the fantastic toy of a well-travelled German princeling. In the event, the 'volcano' of Wörlitz was only primed and exploded on a few special occasions before falling into disrepair in the mid-nineteenth century. (In the later twentieth century, it was stoked only with a few old tyres and was finally condemned in 1983 when part of its weathered stonework collapsed and killed a visitor. It has since, however, been privately renovated and reignited.)[85]

82. For a fuller discussion of the June 1794 eruption of Vesuvius, see chapter 5 in the present volume.
83. On the Wörlitz 'Vesuvius', see the excellent exposition in von der Thüsen, *Schönheit und Schrecken der Vulkane*, p.111-26.
84. Jean-Pierre Houël, *Voyage pittoresque des isles de Sicile, de Malte et de Lipari*, 4 vols (Paris, Imprimerie de Monsieur, 1782-1787).
85. See Andrew Curry, 'That time a German prince built an artificial volcano', *Smithsonian.com* (29 August 2012), https://www.smithsonianmag.com/history/that-time-a-german-prince-built-an-artificial-volcano-19001853/ (last accessed

In essence, what Leopold's monumentally fake Vesuvius achieved on a rather grand scale was little or no different from other eighteenth-century simulations of the volcanic sublime, that is, the domestication of the volcano's awful and unimaginable power. In this sense the Wörlitz volcano chimes perfectly with the smaller-scale replications of volcanic eruptions in the period, such as those famously staged by Humphry Davy in the early years of the nineteenth century. Davy's 1805 public lectures on geology culminated in two spectacular expositions of volcanism, combining the chemist's verbal pyrotechnics with an artificial volcano constructed in the theatre of the Royal Institution. Its eruption was preceded by a reminder of the ineffability of the volcanic sublime: 'when the subterranean fires are displayed in their full energy, when they burst forth from the interior of the earth desolating and destroying, then the effect must be beyond all comparison the most awful and sublime of the phenomena belonging to our globe.'[86] Davy's miniature volcano, modelled in clay, was then made to erupt spectacularly by adding water to a mixture of volatile metallic compounds held in its crater. It thereby provided confirmation of Davy's somewhat retrograde chemical understanding of volcanism; more importantly, it produced controlled but violent explosions and overflowed with fake red-hot lava, to the 'tumultuous applause and continued cheering' of its theatre audience.[87] It simply did not matter that Davy at this point, like the vast majority of his public, had never seen an active volcano. For his effusive demonstration was, by that time, part of an established tradition of conjuring up and taming the volcanic sublime, less by scientific means than by spectacular visual artifice and optical illusion.

The forerunners of Davy's public displays of volcanic activity were the late-eighteenth-century boulevard spectacles of the panorama, the diorama and the eidophysikon. This last device was a primitive cinematographical apparatus creating moving images of sublime natural events such as 'A Storm and Shipwreck' or the slightly more

10 December 2018); see also Gillian Darley, *Vesuvius: the most famous volcano in the world* (London, 2012), p.1-3, 125-30.

86. From Sir Humphry Davy, *Lectures on geology* (1805), cited in Heringman, 'The style of natural catastrophes', p.104.

87. John Ayrton Paris, *The Life of Sir Humphry Davy* (London, 1830), p.218-19. Cited in Heringman, 'The style of natural catastrophes', p.106. For more on Davy and his scientific showmanship, including stage volcanoes, see Richard Holmes, *The Age of wonder: how the Romantic generation discovered the beauty and terror of science* (London, 2008), p.235-304 (294).

disturbing wintry interlude of 'a View of the Alps, a Woodcutter attacked by Wolves'.[88] Given their *son et lumière* brilliance, volcanoes featured prominently in these 'popular landscape entertainments', as Ann Bermingham calls them.[89] They were especially successful in the urban capitals of Europe, particularly in London and Paris. To give but one example, the painter of sublime landscapes and designer of spectacular theatre sets Philippe de Loutherbourg installed his 'Eidophusikon' in a large town house off Leicester Square in January 1782, playing three times a week to packed audiences for five shillings per person. His dazzling light-show of 'various Imitations of Natural Phenomena' necessarily included the stunning spectacle of 'Sunset, A View near Naples', depicting the glowing, craggy outline of Vesuvius.[90] Yet in an age when pseudo-scientific charlatanry vied with genuine scientific experiments for public attention, early volcanologists like Sir William Hamilton did not disdain similar visual gimmickry in promoting their observations. Indeed, Hamilton accompanied a gift of volcanic samples to the Royal Society in late 1767 with a pictorial device comprising a hand-painted transparency of lava flows from Vesuvius, designed to be back-lit, which clearly prefigures Loutherbourg's visual entertainments in its conception if not its purpose.[91] Yet the principal consequence of the creation of multiple artificial volcanoes across the continent in the late eighteenth century was to underline a pervasive aesthetic belief that the sublime might of volcanoes could be tamed and harnessed by art, indeed by clever artifice. Yet, paradoxically, this very positivist aesthetics acknowledged and, in effect, depended upon a persisting sense that the terrible power of an erupting volcano should remain beyond human imagination and human invention.

The uncanny fascination with lava

One of the abiding fascinations of the volcano for scientist and tourist alike was its red-hot lava streams. As we shall see, these lent

88. See the poster, *Eidophusikon, or Various imitations of natural phenomena, represented by moving pictures, invented and painted by Mr. de Loutherbourg* (London, n.n., 1782).
89. See Ann Bermingham, 'Landscape-o-rama: the exhibition landscape at Somerset House and the rise of popular landscape entertainments', in *Art on the line: the Royal Academy exhibitions at Somerset House, 1780-1836* (New Haven, CT, 2001), p.127-43.
90. *Eidophusikon.*
91. See Hamilton, *Observations on Mount Vesuvius*, p.41; and Constantine, *Fields of fire*, p.39.

themselves to easy comparisons with the products of new industrial forges, as in Bergeret de Grancourt's evocation in 1774 of rock 'rendered liquid which runs for long distances like molten lead', or James Smith's description of Vesuvian lava as 'exactly like melted iron poured from a furnace'.[92] (A further, perhaps unintentional, domestication of the volcanic sublime relates to the classification of submarine extrusions as 'pudding lavas'.) Yet lava also resisted these straightforward assimilations to industrial or culinary processes. Especially in its glowing molten state, it captivated all those who encountered it, offering the most perturbing and oxymoronic of volcanic objects to the eighteenth-century observer. The Austrian painter Michael Wutky's 1779 piece *Die Spitze des Vesuvs beim Ausbruch* (Figure 4) perhaps best captures the dizzying sensation of witnessing lava flows up close. The contrast of the brilliant smoking outpourings of the volcano with the cool moonlit sea is reduced to a secondary effect here. Most of the canvas is taken up with enormous cataracts of orange-red lava, blinding the viewer to the dull masses of rock in the foreground and silhouetting tiny gesticulating figures on two uneven ridges of the mountain. The blazing torrents of lava pour confusingly over a flat summit, running down both sides of a dormant crater in the middle ground, before spilling off the painting at its bottom edges. As Luiza Palanciuc has pointed out, lava represents a disconcerting excess of matter, an inexhaustible overflowing of volcanic material which spills and expands in all directions, denying the viewer a single reassuring vanishing point.[93] Yet what is even more disturbing about Wutky's lava is its sheer mobility and liquidity. Its verbal equivalent is Hamilton's rivers of lava compared to the rapidly flowing Severn or Thames, or his 'fountain of liquid fire'.[94] Volcanic lava thus constitutes the ultimate elemental oxymoron: fire as water, rock as fluid. As such, it inevitably intrigues and unsettles, giving rise to elaborate attempts to describe and classify it in the texts of, say, Della Torre, Lalande or Richard.[95]

92. Bergeret de Grancourt, *Bergeret et Fragonard*, p.292; Smith, *A Sketch of a tour on the continent*, vol.2, p.114.
93. Luiza Palanciuc, 'Figures de l'éloignement et paysages volcaniques dans la peinture européenne', in *L'Invention du paysage volcanique*, ed. D. Bertrand, p.187-98 (193-94).
94. Hamilton, *Observations on Mount Vesuvius*, p.10-11, 27.
95. Della Torre, *Storia e fenomeni del Vesuvio*, p.97-98, 112-14; Lalande, *Voyage d'un François*, vol.7, p.165, 189-206; Richard, *Description historique et critique*, vol.4, p.412-13.

Figure 4: Michael Wutky, *Die Spitze des Vesuvs beim Ausbruch* (1779). Vienna, Gemäldegalerie der Akademie der bildenden Künste.

If not scientifically then at least aesthetically, volcanic lava appears to inspire strong feelings of the 'uncanny' in its eighteenth-century observers. The 'uncanny' should be understood here in the Freudian sense of the all-too-familiar (rock) returning in disturbingly unfamiliar or foreign guise (lava).[96] The uncanniness of volcanic lava has little to do with Freudian mechanisms of repression, but it does spring from the same deep 'intellectual uncertainty' as to its real nature, as is also occasioned by other classic objects of the Freudian uncanny (automata, waxworks, etc.).[97] More specifically, it is rock, the very epitome of solidity, which is of a sudden rendered liquid, and so is at once the same substance and apparently another altogether. It instils in its eighteenth-century viewers atavistic doubts of two kinds. Firstly, it suggests that the rationalist worldview of mechanistic Cartesian animals and a clockwork Newtonian universe is fragile and imperilled by undefined irrational forces (the Surrealists' molten clocks play on precisely the same ambivalence).[98] Secondly, it awakens an animist

96. Sigmund Freud, 'The "uncanny"' (1919), in *The Standard edition of the complete psychological works of Sigmund Freud*, 24 vols (London, 1953-1974), vol.17, p.217-52.
97. Freud, 'The "uncanny"', vol.17, p.233.
98. On the eighteenth-century 'invention' of the 'uncanny' as an unintended

fear that what appears to be alive is in fact lifeless, mechanical or sculpted in stone, or vice versa that what appears to be inanimate may spring into life at any moment. Thus hard rock dissolves and runs, the unmovable moves, and brings certain unquestioned assumptions into doubt as to the true nature of things (Emma Hamilton's famous 'Attitudes', or stunning poses struck from antiquity, complement the volcanic lava's uncanniness in the way in which they appear suddenly to freeze soft flesh into unfeeling stone or, conversely, bring marble statues to glowing life).[99] The uncanny, then, adds a further dimension to the fascination exerted by the volcanic sublime.

Industrial volcanoes

Materially, other scientific and social inventions in eighteenth-century Europe instrumentalised a volcanic aesthetic, intimating that the volcano's awful power could be assimilated, if not always completely mastered, by human activity. Chief among these were the industrial forge and furnace. Technological advances in heavy industry in the later eighteenth century appeared to realise perfectly Ernst Cassirer's broad definition of the Enlightenment as the passage from a mythological vision of the world to a scientific one.[100] What else, after all, were these huge furnaces and ironworks but the industrial actualisation of the Cyclops' gigantic forge under Etna, or that of Hephaestus on Olympus? Had not man now mastered the Promethean gift of fire so thoroughly that tales of Vulcan's metal-working prowess were obsolete? The industrial forge had, in other words, replaced the mythological forge.

The close observation of the play of fire and shadow in contemporary ironworks and smithies had already schooled artists such as Joseph Wright of Derby or the Swedish artist Pehr Hilleström. In their studies of white-hot and red-hot furnaces, they deploy important elements of volcanic art: the searing light at the heart of the forge-oven, the orangey haloes and billowing smoke clouds as well as the shadowy workers or darkly silhouetted observers set against the fierce yellows and reds of flame or molten metal. Here the volcanic sublime is

---

consequence of the troubling, accelerated rationalisation of society, see Terry Castle, *The Female thermometer: eighteenth-century culture and the invention of the uncanny* (Oxford, 1995).

99. See Constantine, *Fields of fire*, p.81, 163-66, 169-70.

100. Ernst Cassirer, *The Philosophy of Enlightenment* (Princeton, NJ, 1951).

transformed into an industrial sublime.[101] And this process is confirmed in the very comparisons used by observers of active volcanoes or volcanic debris in the eighteenth century. Thus Faujas de Saint-Fond, a royal commissioner for factories and armouries, likens the sublimation of iron crystals in Auvergne basalts to the workings of steel refineries that he had visited in the Dauphiné, confessing his amazement that 'fire set to work by the hand of man produced a similar effect to that of nature'.[102] In the same text, Faujas also mentions the glass-blowing factory at Roanne, while the more famous Saint-Gobain glassworks near Paris were the site of controlled experiments carried out by their director Pierre Deslaunay-Deslandes to determine whether water introduced suddenly into a working furnace would cause an explosion akin to volcanic eruptions.[103] Joseph Wright had, of course, seen and painted blast furnaces in action before he visited Vesuvius in 1774, and he had also completed sketches of Midlands glasshouses in the early 1770s.[104] Similarly, the operations of both glass-blowing and iron-smelting remain central points of comparison throughout the important volcanological writings of Sir William Hamilton. His very first report to the Royal Society, detailing the March 1766 eruption of Vesuvius, compares the streaming lava to 'a river of red hot and liquid metal, such as we see in the glass-houses', while almost thirty years later the deafening noise of the tremendous eruption of June 1794 reminds him of nothing more than the roar 'produced by the action of the enormous bellows of the Carron iron foundery in Scotland', a modern forge that Hamilton had seen first-hand ten years earlier.[105]

Nor did the volcanic sublime inspire exclusively empirical references to heavy industry. The neo-classical edifices of the utopian French architects Claude-Nicolas Ledoux and Etienne-Louis Boullée also drew directly on volcanic imagery. Ledoux's vision of a clinically geometric cannon foundry at Chaux in eastern France posted four pyramidal chimneys at the corners of his imaginary ironworks, three of which jet blazing fire into the sky. Boullée's grandiose plans for

101. See Thüsen, *Schönheit und Schrecken der Vulkane*, p.84-92.

102. Barthélemy Faujas de Saint-Fond, *Minéralogie des volcans, ou Description de toutes les substances produites ou rejetées par des feux souterrains* (Paris, Cuchet, 1784), p.233.

103. Faujas de Saint-Fond, *Minéralogie des volcans*, p.230-33; for the reference to Deslaunay-Deslandes, see Faujas de Saint-Fond, *Recherches sur les volcans éteints*, p.176-77.

104. See Nicolson, *Joseph Wright of Derby*, vol.1, p.158-69.

105. Hamilton, *Observations on Mount Vesuvius*, p.6; Hamilton, 'An account of the late eruption of Mount Vesuvius', p.78.

a cenotaph borrow the brooding conical form of a volcano while his project for a lighthouse has a colossal smoking tower rise from a rectangular neo-classical base set against swirling clouds. In effect, these neo-classically simplified evocations of the volcanic sublime recall earlier scientific conceptions of the volcano as telluric chimneys and vents. They constitute as much an architectural, as an industrial, sublime; being as much about monumental grandeur as they are about volcanic energy.[106]

Nonetheless, it is worth pointing out that, while heavy industry and utopian architecture provided the aesthetic means to domesticate volcanism, so volcanoes ennobled and aggrandised contemporary industrial processes and neo-classical draughtsmanship. Volcanism quite simply fired the imagination of late-eighteenth-century industrialists, entrepreneurs, artists and architects to conceive of industrial complexes of volcanic proportion and scale. This is significant because the reality of contemporary iron forges and glasshouses was markedly different. The average blast furnace in the late eighteenth century employed between seven and eight workers. And, while the spread of new coke-fired furnaces in the 1770s freed iron smelting from its dependence on local wood supplies, and so increased production in places as diverse as the English Midlands, Liège, Parma and Prussia, the new ironworks remained the exception and not the rule of industrial practice.[107] In fact, industrialisation was chequered and regional; centralised, factory-based, mechanised production was more or less non-existent, and would only become a distinctive feature of socio-economics in the early nineteenth century. In an age better described as one of 'proto-industrialization' or of an '*industrious* revolution',[108] in which urban entrepreneurs and wealthy noble landowners intensified

106. See Claude-Nicolas Ledoux, 'La ville de Chaux, cité idéale de Claude-Nicolas Ledoux', *Passerelle(s)*, http://passerelles.bnf.fr/albums/chaux/index.htm (last accessed 10 December 2018), especially plate 11: 'La forge à canons'; on Etienne-Louis Boullée, see the online exhibition dedicated to him at the Bibliothèque nationale de France: http://expositions.bnf.fr/boullee/arret/d4/index.htm (last accessed 10 December 2018), especially his 'Plaque 7: cénotaphe de Newton, no. 14, élévation géométrale' and 'Plaque 31: fanal à étages, élévation géométrale'. See also Thüsen, *Schönheit und Schrecken der Vulkane*, p.92-93.

107. See Jeremy Black, *Eighteenth-century Europe*, 2nd edn (Basingstoke, 1999), p.52, 56.

108. See *European proto-industrialization*, ed. Sheilagh C. Ogilvie and Markus Cerman (Cambridge, 1996); Jan de Vries, *The Industrious revolution: consumer behavior and the household economy, 1650 to the present* (Cambridge, 2008).

and extended existing rural, domestic production for the growing consumer market, the volcano and its fiery rivers of molten matter allowed those investing in heavy coal-fuelled industries to dream big.

It clearly helped that the volcano had an archetypal form, an instantly recognisable outline (indeed Vesuvius was one of the first objects of Emma Hamilton's drawing lessons in 1786).[109] In the latter years of the eighteenth century and the early years of the nineteenth century, it thus began to figure more frequently in educational texts for children, such as in Friedrich Justin Bertuch's series of *Bilderbuch für Kinder* (1795-1824), the third volume of which features several illustrations of Vesuvius jetting flames, lava and ash as well as interior views of its crater in 1751 and 1775. The fourth volume of Bertuch's series depicts Iceland as a violently eruptive country, with Hekla and another unnamed volcanic peak exploding in the background as Geysir thunders water skywards in the foreground.[110] However, before the spread of cheaper print technologies such as lithography in the early to mid-nineteenth century, the prohibitive price of quality illustrations for volcanological texts such as Hamilton's *Campi Phlegraei* (1776)[111] or deluxe travel narratives such as Saint-Non's *Voyage pittoresque* (1781-1786) meant that the volcano was constructed just as much verbally as it was visually. Hamilton's text tellingly has only one map and informed cross-sections of volcanoes are rare in the period, even if Faujas de Saint-Fond and Lazzaro Spallanzani use cut-away images of volcanic vents in some of their work.[112] In other words, the visually dominant paradigm of patterning, key to much nineteenth-century geology, is almost completely absent from eighteenth-century representations of volcanoes. They are presented primarily via the classical

109. Angelica Goodden, *Miss Angel: the art and world of Angelica Kaufman, eighteenth-century icon* (London, 2006), p.210.

110. Friedrich Justin Bertuch, *Bilderbuch für Kinder*, 12 vols (Weimar, 1795-1824), vol.3 (1798), p.151-62, and vol.4 (1802), p.395-98. The third volume is bilingual, German–French; the fourth quadrilingual, German–French–English–Italian. On volcanic Iceland, see chapter 6 in the present volume.

111. In addition to Hamilton's texts published thus far in the Royal Society's *Philosophical transactions*, this bilingual (English–French) work contains fifty-eight paintings by Pietro Fabris with commentary by the author. On Fabris, see Carlo Knight, 'Sir William Hamilton's *Campi Phlegraei* and the artistic contribution of Peter Fabris', in *Oxford, China and Italy: writings in honour of Sir Harold Acton* (London, 1984), p.192-208.

112. See Rudwick, *The New science of geology*, ch.5: 'The emergence of a visual language for geological science 1760-1840', p.149-95. On Faujas de Saint-Fond and Spallanzani, see chapter 2 in the present volume.

topos of analogy. As such, the inexpressible force of the erupting volcano is nonetheless expressed discursively, and, what is more, it is expressed as a correlative or equivalent of objects possessing much less impressive power or having far more mundane qualities.

This is another facet to the volcano's domestication in the latter part of the eighteenth century and is exemplified by two contemporary accounts of the famous 1779 eruption of Vesuvius. These are Sir William Hamilton's *Supplement to the 'Campi Phlegraei'* and the report given in Simon Linguet's widely read *Annales politiques, civiles et littéraires.* Linguet explains the physics behind the streams of lava spilling over the crater rim by 'precisely the same principle that makes *milk*, *coffee*, *chocolate* and *soap*, and all fatty liqueurs, boil over when they are heated to a certain degree'.[113] In the same vein, Hamilton makes what he calls the 'mean comparison' of aerated lava to 'a rich Parmesan Cheese'; glassy pumice shards resemble 'our common glass bottles' while volcanic bombs are found 'in the shape of a barley corn'.[114] The dreadful might of the volcano is thereby diminished, and, in these analogies with household objects and common foodstuffs, its domestication becomes quite literal. And since parmesan cheese and hot milk are manifestly lacking in sublimity, there is a further important consequence to these analogous configurations of the volcanic sublime: a slow but certain shifting of the locus of the sublime from the awful volcano to its awed, but undeterred, observer and explorer.

From the sublime spectacle to the sublime spectator

This transfer of sublimity from the volcano to its observer-explorer is above all accomplished by what Noah Heringman calls 'eruption narratives', of which Hamilton's *Supplement* of 1779 is a classic instance.[115] The narrator begins by establishing his credentials as a no-nonsense empiricist, eschewing any lyrical rendering of the eruption in favour of 'matters of fact, in plain prose'.[116] (In the event, this is more than a little disingenuous, since Hamilton enlivens his account with a number of typically sublime tableaux.) Nonetheless, the authority of

113. Simon-Nicolas Henri Linguet, 'Italie: éruption du Vésuve du 8 aout 1779', in *Annales politiques, civiles et littéraires du dix-huitième siècle*, 19 vols (London, Brussels and Paris, n.n., 1777-1792), vol.7, p.26-42 (34), original emphasis.
114. Hamilton, *Supplement to the 'Campi Phlegraei'*, p.26-27.
115. Heringman, 'The style of natural catastrophes', p.124.
116. Hamilton, *Supplement to the 'Campi Phlegraei'*, p.5.

his text rests very much on his status as a privileged and informed eyewitness to the eruption. On this his fifty-ninth ascent of the volcano in twenty-five years' residence at Naples, he relates only what either 'his own inspection' or unimpeachable witnesses afford him by way of evidence.[117] Hence his text is dominated by the physical presence of the amateur volcanologist in the field – first-person scientific testimony which is seconded in Hamilton's case by the aesthetic discernment of the world-famous collector of antiquities as well as by the political acuity of the seasoned diplomat. In other words, it is a text that rests as much on the *ethos* or character of its author as it does on the *logos* or rational validity of its arguments.

What is more, in Hamilton's account the scientific observer is also fashioned as a narrative 'hero' figure, a 'knight' on a disinterested, enlightenment quest for knowledge, pitting himself time and again against the terrible might of the volcano to achieve his ends. He relates how, for instance, in May 1779, in the company of a certain Mr Bowdler from Bath and led as ever by his fearless, one-eyed guide Bartolomeo, he strode across the thin crust of a huge lava flow, some fifty to sixty feet wide, in order to escape the choking, sulphurous fumes which had suddenly rolled down on his party from the summit.[118] In this way the intrepid savant provides a second focus for the reader's appreciation of the volcanic sublime, and, as the narrative unfolds, the undaunted scientific observer comes to displace the volcano as the locus of sublimity. This is explicitly confirmed in Hester Piozzi's reflections on her time in Naples, published ten years after Hamilton's *Supplement*, in which she remarks admiringly that 'Sir William Hamilton's courage, learning and perfect skill in these matters, is [*sic*] more people's theme here than the Volcano itself.'[119]

So in Hamilton's volcanological narratives, we find an important articulation of the shift from a Burkean conceptualisation of the sublime to a Kantian one. The sublime is no longer invested in its 'terrible' natural objects (such as the volcano), with the human mind only vicariously 'claiming to itself some part of the dignity and importance of the things which it contemplates', as Burke put it.[120] Now the sublime resides in man himself, for, in Kant's words, 'it is not so much the object but rather a cast of mind in appreciating it that we

117. Hamilton, *Supplement to the 'Campi Phlegraei'*, p.1.
118. Hamilton, *Supplement to the 'Campi Phlegraei'*, p.3-4.
119. Piozzi, *Observations and reflections*, p.351.
120. Burke, *A Philosophical enquiry*, p.50-51.

have to adjudge as sublime.'[121] Far from contributing only marginally to an apprehension of the sublime, the amateur volcanologist becomes its embodiment, incarnating a Kantian moral grandeur which allows him to discover in himself 'a superiority over nature that founds an altogether different kind of self-preservation from that which may be challenged and endangered by external natural forces'.[122] And he is consolidated in this exalted role by two other cultural factors of the time. Firstly, he becomes eligible to join the cult of the scientist as genius and benefactor of humanity, enshrined in the eulogies and elegies practised assiduously by a growing network of major and minor academies across Europe.[123] Secondly, he benefits from a virile identification of the sublime with displays of a specifically masculine valour. Such a gendered reading of the male mind as sublime was already present in Kant's *Beobachtungen* (1764), but it was championed in public discourse too, as in the academician Antoine-Léonard Thomas's *Essai sur les femmes* (1772) which distinguishes women's vivid, excitable but weak imagination from that of men, 'always active and exposed to storms', whose poetic bent is energised 'on mountain tops, on the brink of volcanoes, on high seas, on battlefields and amidst ruins'.[124] In England it was, rather oddly, the youngest sister of Sir Joshua Reynolds who confirmed this exclusive identification of sublimity with masculinity. Frances Reynolds' *Enquiry concerning the principles of taste, and the origin of our ideas of beauty* (1785) schematically genders the beautiful and the sublime, characterising feminine commonplaces of placidity (lamb, dove, lake, valley) as beautiful while their vigorous masculine contraries (steed, eagle, ocean, mountain) represent sublimity. 'It is the feminine character', she writes, 'that is the sweetest, the most interesting, image of beauty; the masculine partakes of the sublime.'[125]

If we turn again now to Linguet's account of the same eruption of Vesuvius in August 1779, we find that it broadly corroborates this transfer of the volcanic sublime to the sublime volcanologist, with the creation of a similar, bold savant-hero. Yet, as Linguet is clearly

121. Kant, *Kritik der Urteilskraft*, p.178.

122. Kant, *Kritik der Urteilskraft*, p.186.

123. See, for example, Jean-Claude Bonnet, *Naissance du Panthéon: essai sur le culte des grands hommes* (Paris, 1998).

124. Antoine-Léonard Thomas, *Essai sur le caractère, les mœurs et l'esprit des femmes dans les différens siècles* (Paris, Moutard, 1772), p.112-13.

125. Frances Reynolds, *Enquiry concerning the principles of taste, and the origin of our ideas of beauty, etc.* (London, Baker and Galabin, 1785), p.29. See also p.17-19 for the identification of the sublime with 'terror', 'power' and 'genius'.

not able to fulfil this role himself, it is taken in his text by Giovanni Maria Della Torre, the pioneering explorer of Vesuvius in the 1750s. Having witnessed the major eruption of the volcano in 1751 with what Linguet qualifies as 'an attention and courage especially rare among scholars', Della Torre is cited in his own words, recounting a subsequent experiment which, Linguet promises, will make the reader shudder to hear it.[126] Thus in October 1752, in order to measure the depth of the boiling crater, Della Torre relates how he inched out 'over the abyss' on an outcrop of rock 'about twelve feet in length'. He then drops a stone into the roiling pit beneath him, but, before he has timed more than two thirds of its fall, he is suddenly enveloped in suffocating smoke and has to dive to reach clean air at the other end of the thin promontory. Apparently beaten but unbowed, the savant registers his moral victory over the spluttering volcano by deducing the full depth of the crater from the time taken for the stone to fall two thirds of the distance to the heaving lava beneath (calculated at 543 feet). It is perhaps more telling still that Della Torre refers to the bubbling crater as 'a vast cauldron full of molten crystal', a domesticating metaphor that allows Linguet to develop his analogy of overflowing lava with boiling pots of coffee, milk or chocolate.[127] A year earlier, confronting an advancing lava flow, Della Torre anticipates Hamilton's tendency to collapse the distance between himself and the terrible object of his observations – 'I kept to a distance of thirteen or fourteen feet' – all the more surely to appropriate the sublimity of the irresistible tide of molten rock. He continues: 'From that distance, I felt a considerable heat, but far from troubling me, on the contrary it gave me strength and vigour.'[128] Infused with the force of the volcano, arrogating its sublime might to himself, the volcanologist employs it to master the erupting volcano through his coolly taken assessments of the movement, consistency and colour of its lava flows.

In this sense, the uniquely intense heat generated by the eruption turns on its head another commonplace of scientific and travel literature of the time, namely, the widely held climate theory whereby the warm South induces languor, indolence and effeminacy in its residents while the North braces and invigorates those who are daily

126. Linguet, 'Italie: éruption du Vésuve du 8 aout 1779', p.35. He cites Della Torre from the French translation of his major work on Vesuvius, *Histoire et phénomènes du Vésuve* (Paris, Hérissant, 1760).

127. Linguet, 'Italie: éruption du Vésuve du 8 aout 1779', p.35-36.

128. Linguet, 'Italie: éruption du Vésuve du 8 aout 1779', p.37.

exposed to its chill.[129] Here, in contrast, the volcano, as the hyperbole of fierce heat, energises and enthuses anyone who is prepared to approach it, first among whom are our undaunted earth scientists. Wherever he is from – Hamilton is from the bitterly cold North, Della Torre from the torrid South – the intrepid savant wrests a degree of sublimity from the volcano each time he faces down its terrible dangers, and, since the sublime is usually gendered as masculine, in the process the volcano imparts increased virility to those who 'conquer' it.

However, one of the problems with the volcano providing a natural site for the creation of sublime savants and scientist-heroes is that the science itself can often be eclipsed in the 'eruption narrative' by the thrill of the tale. As a result, the volcanic sublime moves within the compass of non-specialists at precisely the period when more and more non-scientific tourists were coming to scramble on the slopes of the volcano and experience its sublimity first-hand. The fact that the night-time spectacle of an eruption was increasingly an integral part of confronting the sublime volcano suggests that, for these tourists at least, the prime motivation for their volcanic adventure was not scientific but aesthetic. Certainly, in many of the later travel narratives of the eighteenth century which feature an encounter with Vesuvius, the affective takes over from the objective. The overall outcome is, however, strikingly the same: to effect the transfer of the sublime from the spectacle to the spectator, from the volcano to its explorer.

Jean-Marie Roland de La Platière is exemplary on this count. He visited Vesuvius three times in 1776 and 1777 and published an account of his experiences three years later in his *Lettres écrites de Suisse, d'Italie, de Sicile et de Malthe,* addressing the text more or less openly to his then fiancée, Marie-Jeanne Phlipon. Here he is skipping fearlessly like Hamilton over lava flows – 'the least slip, a false step would have been fatal' – confronting the awesome might of the erupting volcano:

129. See Chard, *Pleasure and guilt on the Grand Tour,* p.119-25. On climate theories more generally, see Jean-Patrice Courtois, 'Le climat chez Montesquieu et Rousseau', in *L'Evénement climatique et ses représentations (XVIIe-XIXe siècle),* ed. Emmanuel Le Roy Ladurie (Paris, 2007), p.157-80. Déodat de Dolomieu applied Montesquieu's prevailing climate theories to the volcanic Aeolian Islands when he explored them in the summer of 1781. The local difference in vigour and industriousness of the populations on Lipari and Salina was explained in terms of the former being invigorated by their proximity to active Vulcano while the latter basked in relative calm and were cheerfully indolent as a consequence. See Cocco, *Watching Vesuvius,* p.209-18.

> The noise had greatly increased; the mountain shook; stones flew right up to me, and I had to judge their flight in the air to avoid them [...]. Finally the flames, the swirling clouds of smoke which reached right up to me, even though I was not downwind of them, this terrifying spewing of stones and lava which fell right at my feet; [...] all of this was of a character to inspire terror.[130]

Rhetorically noteworthy is the repeated construction 'right up to me', 'right at my feet', signalling that the volcano is actively attacking Roland, but, like the most intrepid savant-hero, he defies its renewed assaults, and in so doing appropriates its sublimity. Again, as for Della Torre before him, once mastered, the heat of the volcano no longer threatens Roland but on the contrary empowers and impassions him: 'I dream only of the volcano, Vesuvius' heat inflames my love for it.'[131] It inspires him less to scientific musings than to seek out captivating nocturnal vistas of its flaming cone and to commission three tableaux of the volcano's 'majestic and ravishing effects'.[132] It is instructive that, for this new breed of volcanic adventurer, it is only on Roland's third and final visit to the crater that he is composed enough to collect some rock samples. In other words, it is only belatedly that the scientist picks up some crumbs from the sensationalist's feast. With the generation of thrill-seeking tourists of the 1770s and 1780s, what was formerly a scientific venture of discovery becomes primarily a cultural and aesthetic journey of self-discovery, not to say, of volcanic self-fashioning.

Yet, as subsequent chapters will show, this empowering self-realisation in the face of the volcano was not exclusively aesthetic in character. It was also informed by major intellectual upheavals earlier in the century and would, in turn, strongly mark the revolutionary movements at the century's close.

130. Roland de La Platière, *Lettres écrites de Suisse*, vol.2, p.243, 235-36.
131. Roland de La Platière, *Lettres écrites de Suisse*, vol.4, p.39.
132. Roland de La Platière, *Lettres écrites de Suisse*, vol.4, p.40.

4. More heat than light? Natural philosophies of volcanism

Jonathan Israel has convincingly charted the rise of dangerous schools of atheism and materialism in late-seventeenth-century Europe.[1] As part of the Church-led counter-offensive against these forms of irreligious thinking, one branch of Christian theology looked to the new earth sciences for intellectual succour. This was physico-theology. It recast biblical cosmogony in geophysical terms, seeking and finding in the natural world, and especially in geology, its irrefutable proofs for intelligent design. According to this theological reinterpretation of Earth history, the Flood could be explained and verified in terms of Cartesian mechanics while the activity of an 'Almighty designer' was to be induced from various exercises in Newtonian empiricism.[2] The culmination of this thinking was perhaps the Swiss scholar Johann Jakob Scheuchzer's *Physica sacra* (1731-1735), which systematically correlated contemporary science to biblical verses. For instance, Scheuchzer's own hygrometrical measurements proved that the Alps received twice as much rainfall as continental lowlands, making it the 'water reservoir of all Europe'. This fact was not, however, to be treated as a simple 'necessity of nature' but as a providential blessing granted by 'a particular disposition of God's wisdom'.[3] Its high-mountain

1. Jonathan I. Israel, *Radical Enlightenment: philosophy and the making of modernity 1650-1750* (Oxford, 2001).
2. This term comes from both Isaac Newton's *Mathematical principles of natural philosophy* (1729) and William Wollaston's *Religion of nature delineated* (1724), insofar as both try to reconcile mathematical proofs with the physical world as evidence of intelligent design. Cited in Jonathan I. Israel, *Enlightenment contested: philosophy, modernity and the emancipation of man 1670-1752* (Oxford, 2006), p.208.
3. Johann Jakob Scheuchzer, *Physica sacra*, 4 vols (Augsburg and Ulm, Christian Ulrich Wagner, 1731-1735); citations here are taken from the French edition,

topography transformed Switzerland, in physico-theological terms, into a new well-watered Canaan. In this much, its lot was blessed after the biblical Flood, unlike the Dutch lowlands reclaimed by the sea or the kingdoms of Sicily and Naples in which 'whole towns [have been] ruined by volcanoes'.[4]

The most influential treatise of physico-theology also drew on an early exploration of the Alps. But it was equally, if not more, informed by the millenarian visions of late-seventeenth-century England. This was Thomas Burnet's *Telluris theoria sacra* or *Sacred theory of the Earth* (Latin 1680; English 1684), which ran to at least seven re-editions before 1759. It might best be described as a speculative cosmogony which frames Earth history between the biblical cataclysms of a past Great Flood and a future Great Fire. Yet the geology of both these apocalyptic events is that of seventeenth-century European science. On the one hand, they conform to prevailing Cartesian geo-mechanics. On the other, they are informed by Kircher's conception of a globe traversed internally by huge channels of roaring fire and turbulent bodies of water. As we have seen, the subterranean fires breach the Earth's crust to form volcanoes. The masses of water fill huge dendritic cavities and, under internal pressure of fire and air, force water out through systems of aquifers in ocean basins and as mountain springs which swell to rivers.[5] For Burnet, the biblical Flood constitutes the archetype of these latter geological processes, releasing unimaginable torrents of internal water which hollowed out the Earth and ravaged its surface with collapsed and buckled structures. The Flood then destroyed the seamless paradise of Eden before sin entered the world; it recast the globe in the deformed image of fallen mankind. In this much, Burnet exhibited a classical revulsion at the bulging or shattered mountain scenery that the Flood was thought to have left in its wake: 'how barren, how desolate, how naked are they.' Mountains stand as the admonishing 'great ruines' of a post-diluvian world.[6]

In Burnet's version of physico-theology, volcanoes play a dual role, at once physical and metaphysical, natural and supernatural.

Physique sacrée, ou Histoire naturelle de la Bible, 8 vols (Amsterdam, Pierre Schenk; Pierre Mortier, 1732-1737), vol.7, p.161-62, 61.

4. Scheuchzer, *Physique sacrée*, vol.7, p.82-83.
5. See the plate 'Systema ideale quo exprimitur aquarum per canales hydrogogos subterraneos', in Kircher, *Mundus subterraneus*, p.174-75.
6. Thomas Burnet, *The Sacred theory of the Earth*, 3rd edn (London, Walter Kettilby, 1697), p.95-97.

Empirically, they stand with caves, grottoes, earthquakes and mines as proof of the hollowness of the post-diluvian Earth emptied of its waters. In a language redolent of the imagery of Kircher and, to a lesser extent, Lémery, volcanic craters lead like a 'Funnel' to subterranean 'Magazine[s] of combustible Matter', the explosion of which may be intensified by infiltrating seawaters.[7] Burnet bases his presentation of volcanism on the historical examples of eruptions of Vesuvius, Etna, 'Strombolo', Hekla in Iceland and the Aeolian Islands off Sicily.[8] He goes so far as to cite the mathematical calculations of Giovanni Alfonso Borelli regarding the immense quantities of lava ejected from Etna in 1669 – estimated as a body of molten matter that would have stretched four times round the Earth in a layer three feet deep.[9] Nor does he omit the powerful underwater volcanism of the Azores, forcing up the island of St Michel from 'a depth of an hundred and sixty fathoms' below the surface of the sea.[10]

Yet these historical-empirical volcanoes remain subordinate to the theology governing Burnet's conception of the Earth. They are the natural agents of a millenarian vision, ranking as secondary causes set to work in the world by an inscrutable primary cause, the all-powerful will of God. The volcanic activity known to human history is but a fizzling foretaste of its providential function as the 'natural Causes and Materials' of the 'Last Fire'.[11] As Burnet puts it, these volcanoes have hitherto been only the 'lesser Essays or preludes to the general Fire; set on purpose by Providence to keep us awake, and to mind us continually, and forewarn us of what we are to expect at last'.[12] Hence the very material characteristics of volcanoes carry a spiritual significance: their seeming inexhaustibility means they will be active still at the final Great Fire; their dispersal around the known globe is providentially conceived to ensure a 'universal combustion'; and their marked presence in southern Italy suggests that the apocalyptic conflagration will kindle at the heart of Christendom. In Burnet's fervently pious imagination, the singular eruptions of a Vesuvius or an Etna are conjoined and conflated into a fiery apocalypse consuming

7. Burnet, *The Sacred theory of the Earth*, p.79, 53.
8. Burnet, *The Sacred theory of the Earth*, p.80.
9. Burnet, *The Sacred theory of the Earth*, p.39.
10. Burnet, *The Sacred theory of the Earth*, p.52. In fact, St Michel is one of four overlapping strato-volcanoes rising 1500 metres from a triple tectonic-plate junction on the floor of the North Atlantic.
11. Burnet, *The Sacred theory of the Earth*, p.35.
12. Burnet, *The Sacred theory of the Earth*, p.37.

the whole planet.[13] Earth history is then bookended by two devastating deluges – one of water, one of flame – both breaking from the interior of the Earth. In the last, volcanoes even shed their aura of dreadful destruction, insofar as Burnet also envisages them as providentially primed fireworks announcing Christ's triumphant second coming.[14] Through volcanic fire, nature is ultimately translated into scripture.

Burnet's physico-theology was widely admired in early-eighteenth-century Europe for its skilful synthesis of natural history and revealed religion.[15] A number of texts drawing on Burnet's work laid more emphasis on the benevolence of the divine creation, rather than its monstrous deformation by sinning humanity, let alone its apocalyptic resolution. In France François Fénelon's *Demonstration de l'existence de Dieu, tirée de la connoissance de la nature* (1712) marshalled arguments purporting to show how the bounties of nature corroborated a divine plan. Thus, in his consideration of geological phenomena, he praises the inexhaustible liberality of the Earth, including volcanoes that supply humankind with metallic ore and a source of fire.[16] William Derham's *Physico-theology* published a year later goes even further in calling volcanoes the 'wise contrivances of the Creator, serving great Uses of the Globe, and Ends of God's Government'.[17] While conceding that volcanic eruptions may serve to scourge and chasten local populations given over to wanton sinfulness and so stand as 'Emblems, and Presages of Hell it self', they also vent fire and deadly vapours that would otherwise 'make dismal Havock' among the innocent.[18] In this sense, they are doubly providential. Interestingly, Derham employs a mechanistic geophysics to explain the beneficial actions of these 'fiery Vents' which, he observes, occur frequently in earthquake-ravaged regions, thereby sparing them even worse devastation than that which they actually endure.[19]

13. Burnet, *The Sacred theory of the Earth*, p.39-43.
14. Burnet, *The Sacred theory of the Earth*, p.55-59.
15. On the reception of Burnet's work, see Rhoda Rappaport, *When geologists were historians, 1665-1750* (Ithaca, NY, and London, 1997), p.139-49.
16. François de Salignac de La Mothe Fénelon, *Demonstration de l'existence de Dieu, tirée de la connoissance de la nature, & proportionnée à la foible intelligence des plus simples*, 2nd edn (Paris, Jacques Estienne, 1713), p.21-27, 42-44.
17. William Derham, *Physico-theology, or a Demonstration of the being and attributes of God, from his works of creation* (London, W. Innys, 1713), p.68.
18. Derham, *Physico-theology*, p.69.
19. Derham, *Physico-theology*, p.69-70. Derham is here citing approvingly John Woodward's *An Essay towards a natural history of the Earth* (1695).

Yet this very method of deducing proofs of God's foresight and benevolence from the physical features and events of the natural world – one of the intellectual trumps of physico-theology – also proves to be its principal philosophical weakness. Why, sceptics and freethinkers asked, should these proofs apply only to the cosmic kindness of a Christian God? Might not the argument from design lend itself just as well, if not more, to a *natural*, rather than a revealed, religion, having no need of mystical prophet-interpreters to discern providence at work in Alpine rainfall or volcanic eruptions? The response from Christian apologists – to assimilate nature's 'marvels' to biblical 'miracles' – was far from convincing.[20] It did not stand up to scrutiny to assert that God chose to deploy the secondary causes of nature to supernatural effect only in those instances when a priest said so and a credulous fool agreed. Holy miracles were demystified, debased by the comparison; they became a mere theologising of what would later be called the natural sublime. As a result, in the 1730s and 1740s interpretations of natural phenomena such as volcanoes, propagated and popularised by physico-theology, were subsumed under a more general providential deism which frequently repudiated miracles. Thus Thomas Morgan's *Physico-theology* of 1741, while still generally subscribing to the beliefs of the Church, rejected all notion of miracles, scorning as absurd those who 'talk to Men of a supernatural Light, or of Revelation above Reason'.[21] Or, more tellingly still, the abbé Noël-Antoine Pluche's hugely successful *Le Spectacle de la nature* (1732-1742) abjured the miraculous in favour of the naturally spectacular or the daily wondrous. It is a work in which volcanoes are presented as the providential means of diffusing the tremendous destructive energies of underground chemical explosions so that the volcano 'which one considers to be the scourge of the region in which it is found, by the designs of God, is in fact its salvation'.[22] By the time Buffon published the first volume of his *Histoire naturelle* in 1749, even this figure of the tutelary, intelligent deity could be gently dispensed with. Buffon's account of the slow but consistent convolutions of matter from

20. See Israel, *Radical Enlightenment*, p.456-64.
21. Thomas Morgan, *Physico-theology, or a Philosophical-moral disquisition concerning human nature, free agency, moral government, and divine providence* (London, T. Cox, 1741), p.321.
22. Pluche, *Le Spectacle de la nature*, vol.3 (1735), p.269-70; see also Armand Guilhem, '*Le Spectacle de la nature* ou l'esthétique de la Révélation', *Dix-huitième siècle* 45:1 (2013), p.329-45.

prehistory to the present, subtly divested of divine guidance, implies the workings of an eternally indifferent, self-regulating, materialist nature. The volcano, as we have seen, becomes just a fiery, noisy sideshow in this first iteration of Buffon's earth science.

An anti-clerical volcano

By the time of the Lisbon disaster in November 1755 the Catholic Church's favoured allegorical presentation of the volcano was clearly outdated. Its belching flames reminded fewer and fewer of the faithful of the horrors of hell-fire. The stench of sulphur was analysed as chemical rather than infernal, and the rain of ashes was known to enrich the soil, rather than recall the sinful to symbolic penitence. Only when Vesuvius erupted violently and threatened Naples did the miraculously liquefied blood of San Gennaro offer a potent counter-force to the blood-red lava flows. It provided a Galenic equal and opposite humour to cancel out the volcano's choler and a timely reminder of Christian salvation by the blood of Christ.[23] Yet its supernatural appeal was increasingly limited to the more superstitious lower orders of society. For a convinced Catholic who was nonetheless familiar with the wonders of modern science, a delicate balance had to be struck between the geological and the theological. This was a particularly acute predicament in the territories of the Inquisition.[24] Hence the Spanish university professor Francisco Martínez Moles responds with some circumspection to the Lisbon earthquake in his *Dissertacíon physica* of 1755. He juggles his faith and his science by maintaining that the primary, divine and providential, causes for the disaster cannot be questioned; but the secondary causes of the earthquake, conforming in themselves to the known laws of nature, can and should be investigated. In this context, he feels able to explain 'underground fires' as the common cause of earthquakes and volcanoes, resulting from subterranean canals of water and wind (*dixit* Kircher) admixing, agitating and igniting huge deposits of flammable nitrous, sulphurous and bituminous materials (*dixit* Lémery).[25] Nonetheless,

23. See Marie Viallon, 'Le *buon gigante*', in *Le Vésuve en éruption*, ed. E. Beck Saiello and D. Bertrand, p.41-50 (47).
24. For Naples under the Spanish Habsburgs, see Cocco, *Watching Vesuvius*, p.170-73.
25. Francisco Martínez Moles, *Dissertacíon physica: origen, y formacíon del terremoto, padecido el día primero di noviembre de 1755* (Madrid, Juan de San Martín, 1755), p.9-10; see also Augustín Udías, 'Earthquakes as God's punishment in 17th- and

these secondary causes remain always at the bidding of the Lord and, in the Last Judgement, 'vomiting volcanoes' will be only one of the forces of nature massed against an irredeemably sinning mankind.[26] More interestingly perhaps, in Moles' text, the volcano also appears as a simile for the irrepressible character of the devout Catholic's passion for the natural sciences. He writes that, while his devotion to the true faith 'extinguishes' all others, his enthusiasm for 'physics' only lies dormant and is wont to erupt from time to time.[27] This hints, ever so tentatively, at a shift in the balance of power between the competing religious and scientific drives to explain large-scale geological phenomena.

There may also have been other unintended scientific consequences to the Catholic Church's insistence on mankind's incorrigible sinfulness as a prompt for natural catastrophes. As Lydia Barnett has intriguingly suggested, this implicitly posits humans as agents as well as victims of major geological or environmental disasters and so, by implication, of global climate change.[28] Enemies of the Church certainly seized on and satirised its tendency to extrapolate messages of universal doom from localised accidents of geology. They also recognised that physico-theology and providentialism were similarly prone to globalising abstractions. The most famous example of this comes in Voltaire's *Candide* (1759), with Dr Pangloss's systematically providentialist explanation for the Lisbon disaster. Pangloss is an 'optimist', believing stubbornly in the notion that the world has been conceived by an omnipotent and benevolent deity to contain an 'optimum' of good and thus to be 'the best of all possible worlds'. Hence nothing happens in the world without it having a reasonable and beneficent cause, no matter how appalling the event might seem. Adhering blindly to this conviction through a series of picaresque misadventures, each of which proves his folly, Pangloss and his charge, the young and naïve Candide, are shipwrecked by the tsunami accompanying the Lisbon earthquake. Candide lies injured among the rubble of the city, but Pangloss ignores his cries of pain, absorbed

18th-century Spain', in *Geology and religion: a history of harmony and hostility*, ed. M. Kölbl-Ebert (London, 2009), p.41-48. I am also grateful to Dr Carla Almanza Gálvez for her insightful advice on this text.

26. Martínez Moles, *Dissertacíon physica*, p.28.
27. Martínez Moles, *Dissertacíon physica*, p.3.
28. Lydia Barnett, 'The theology of climate change: sin as agency in the Enlightenment's Anthropocene', *Environmental history* 20 (2015), p.217-37.

in his 'optimist' reasoning that: 'This earthquake is no new thing, [...] the city of Lima in South America experienced the same tremors last year; the same cause, the same effects; there is undoubtedly a trail of sulphur underground all the way from Lima to Lisbon.'[29] Pursuing this line of 'optimist' thought the following day, he consoles the locals among the debris of their city with this deduction: 'All of this is for the very best, for if there is a volcano in Lisbon it could not be anywhere else; for it is impossible that things should not be where they are, for everything that is is good.'[30] Unfortunately for Pangloss and Candide, the Doctor's ratiocinations are overheard by an Inquisitor and the unlucky pair are rounded up to be executed as heretics in expiation of the city's collective sins which had allegedly brought the earthquake upon them all. Or, as Voltaire wryly asserts, it had been 'decided by the University of Coimbra, that the spectacle of a few people burned alive over a slow fire with great ceremony, is an infallible secret means of preventing earthquakes'.[31]

For Voltaire, the Church's abusive interpretation of natural catastrophes was just a further illustration of its inveterate ignorance, intolerance and fanaticism. From the 1760s onwards, other French *philosophes* were to draw on similar examples in their anti-clerical campaigns. The disciple of Buffon and proponent of deep geological time Jean-Baptiste-Claude Delisle de Sales uses a general figure of 'volcanic' violence to condemn religious fanaticism. In his rambling, compendious *De la philosophie de la nature* (1770) he states that 'fanaticism is this blazing torrent that rushes from the volcano's summit and scours the plain in order to lay waste to it, leaving livid traces of destruction everywhere.'[32] In a more conciliatory vein, he adds that, if the Saint Bartholomew's Day massacre of 1572 stands as the acme of past religious intolerance, 'the volcano of fanaticism seems now to have been forever sealed shut amongst us, even if it still throws out a few sparks from time to time.'[33] Simon Linguet, in his decidedly partial *Histoire impartiale des jésuites* (1768), also uses the metaphor of the volcano figuratively. He employs the image of an interconnected

29. Voltaire, *Candide, ou l'Optimisme*, ed. René Pomeau, in *The Complete works of Voltaire*, vol.48 (Oxford, 1980), p.136.
30. Voltaire, *Candide*, p.137.
31. Voltaire, *Candide*, p.138.
32. Jean-Baptiste-Claude Delisle de Sales, *De la philosophie de la nature*, 3 vols (Amsterdam, Arkstée & Merkus, 1770), vol.2, p.2.
33. Delisle de Sales, *De la philosophie de la nature*, vol.2, p.56-57.

web of subterranean fires in order to denounce the conspiratorial ambitions of the Society of Jesus. Their subversive plots are allegedly communicated instantly through a pan-European network of monastic cells in order to foment religious strife and topple thrones; their violence thus erupts discretely like Vesuvius or Etna, but is in fact part of a larger 'underground' design to extend the dominion of their papal master in Rome.[34] Less figuratively, Linguet cites the bloody colonial ravages of one of the first sixteenth-century Jesuit missionaries, François Xavier, as he ranged through the Portuguese East Indies. After leaving a trail of murder, persecution, forced conversion and expropriation among various heathen populations in the Indian Ocean and the Malay peninsula, the Jesuit zealot sailed to the Molucca Islands. There, claims Linguet, 'he made the local inhabitants believe that a volcano, viewed with dread because of the flames it spewed, was one of the vents of Hell. On the strength of this lie, he was said to have baptised 25,000 souls.'[35]

A different form of mordant Voltairean mockery defines the Neapolitan economist and reformer Ferdinando Galiani's attack on his fellow citizens' craven superstitiousness in the face of the 1779 eruption of Vesuvius. Galiani penned a pamphlet that circulated as the eruption continued whose very title ridiculed its subject matter, purporting to be a *Most extremely frightening description of the frightening fright that frightened all of us with the eruption of Vesuvius on the evening of the eighth of August 1779, but (by grace of God) did not last long.*[36] The text knowingly parodies the fantastical clerical accounts of the terrible 1631 eruption but also usurps the voice of the contemporary demagogue Onofrio Galeota. Its relentless bombast and hyperbole mock, and so defuse, the hysteria in the streets and on which the Church was wont to feed. As well as being caustically anti-clerical, however, the short text also aimed its scatter-gun derision at fashionable hold-all theories of volcanism, such as electricity and chemical fermentation. Thus, it maintained, those scientists and scholars who scoured Vesuvius' hot slopes, seeking 'the where, the how, the when and the why', never understand so much as a

34. Simon-Nicolas Henri Linguet, *Histoire impartiale des jésuites, depuis leur établissement jusqu'à leur première expulsion*, 2 vols (n.p., n.n., 1768), vol.1, p.289.
35. Linguet, *Histoire impartiale des jésuites*, vol.1, p.356. The volcano in question is probably Mount Gamalama on Ternate.
36. Ferdinando Galiani, *Spaventosissima descrizione dello spaventoso spavento che ci spaventò tutti coll'eruzione del Vesuvio la sera degli otto d'agosto 1779, ma (per grazia di Dio) durò poco* (Naples, 1825).

'jot' about the mountain.[37] If this might indirectly lampoon Hamilton and his many guests, scrambling among the cooling lava flows, the British diplomat's own accounts of the same volcano's eruptions shared Galiani's scathing anti-clericalism. His report on the paroxysmal eruption of June 1794, for instance, mockingly cites a text of 1631 which had imputed the mountain's 'malevolence' to 'devils'; and his text goes on to pour scorn on the imbecilic nuns of Torre del Greco who warmed their hands on the encroaching lava and were keen only to save their sweetmeats from the flames. Hamilton describes their cloistered minds as being as confined and unfurnished as their convent cells.[38] These anti-clerical comments draw very differently on the volcano as a site of extreme violence or subterranean communication, as fostering fearful superstition or proving scripturally inspired ignorance. What unites them is a denunciation of the Church's persistent refusal to embrace enlightened toleration and reason.

The use of volcanoes to express mocking or vitriolic anti-clerical sentiment was not restricted to texts. Eighteenth-century tableaux depicting erupting volcanoes also contained ciphers of contempt for the clergy. This was particularly the case for paintings representing Vesuvius ablaze, such as Carlo Bonavia's *Eruption of Vesuvius from the slopes of the crater* (1756-1757).[39] Among the usual gesticulating spectators in this picture there are two well-dressed gentlemen prodding a stick into a lava flow and marvelling as it catches fire. They are subtly contrasted with a dwarfed religious procession on the lower slopes that brandishes a large cross ineffectually in the direction of the vivid scarlet-orange summit. Inquisitive empiricism is juxtaposed with unthinking superstition, and the colour tones suggest a literal enlightenment stands in opposition to benighted ignorance. Personal testimonies of visits to an erupting Vesuvius confirm that the practice of throwing stones into lava flows was not uncommon. It was better still if one could get close enough to stab sticks into the molten rock to see them flare. Roland de La Platière does just that in October 1776, recording an initial resistance in the crust of the lava before his stick sank in 'as though in water' and burst into flames.[40]

37. Galiani, *Spaventosissima descrizione*, p.7. See also Cocco, *Watching Vesuvius*, p.231-33.
38. Hamilton, 'An account of the late eruption of Mount Vesuvius', p.73, 86-87.
39. Carlo Bonavia, *Eruption of Vesuvius from the slopes of the crater* (1756-1757), Rome, private collection.
40. Roland de La Platière, *Lettres écrites de Suisse*, vol.2, p.242. Roland was told that

The enlightened community which gathered on the volcano was inclusive too, as news-sheets reported the presence of princess d'Esterhazy, wife of the Austrian ambassador, in one party of natural scientists inspecting the raging lava flows in 1751.[41] The tableaux of Vesuvius erupting by Pierre Jacques Volaire also bear witness to the presence of women in the lee of violently orange jets and rivers of lava. Thus his *Eruption du Vésuve vue depuis l'Atrio del Cavallo* (1785-1786) – a variation on his classic composition of this scene from 1771 – includes a miniature portrait of a woman calmly sketching the roaring eruption, whom Emilie Beck Saiello suggests could be either Hester Lynch Piozzi or Margaret Georgiana Poyntz, Vicountess Spencer.[42] Philipp Jakob Hackert's *Der Ausbruch des Vesuv* (1774) goes further still in portraying a disinterested curiosity regarding the volcano's explosive activity. This nocturnal spectacle dispenses with all religious references to present a group of well-to-do men and their guides by the edge of a gushing lava flow. One man pokes at a pile of volcanic debris while others lean on canes, chatting, as they observe the orange and scarlet torrents of molten rock coursing by them. There is little terror or awe in their poses; if anything the men appear phlegmatically to defy the volcano's red-hot fury. The scene could be one of an iron foundry at night.

Hackert's painting explicitly references the towering spouts of lava that burst from a new vent on the north–north-eastern face of Vesuvius in January 1774. The later throes of this eruption were also witnessed by Joseph Wright of Derby during his brief sojourn in Naples that year. Wright's subsequent output of volcanic tableaux runs the gamut from Hackert-like studies in the industrial sublime to Turner-style abstractions of crags, flames, smoke and ash. Yet his *Eruption of Mount Vesuvius with the procession of St. Janarius's head* (1778) revisits, albeit more darkly, the religious motif of Bonavia's earlier painting. Here, however, the diminutive procession crossing the eerie nocturnal scenery, with their shrivelled holy relic borne aloft, meets with no enlightened alternative. Instead, the futility and fatuity of

---

the lavas of the 1767 eruption had been so incandescently hot that they had melted iron when it was thrust into them (vol.2, p.244).

41. This is reported in the Swiss *Almanach du messager boiteux* of 1751 in its account of the powerful eruption of Vesuvius that year. Cited in René Favier, 'Les éruptions du Vésuve dans la littérature populaire de la seconde moitié du XVIIIe siècle', in *Le Vésuve en éruption*, ed. E. Beck Saiello and D. Bertrand, p.99-109 (106).
42. Pierre Jacques Volaire, *Eruption du Vésuve vue depuis l'Atrio del Cavallo* (1785-1786), Naples, private collection. Beck Saiello, *Pierre Jacques Volaire*, p.163.

their mission are conveyed by the almost infernal orange glow of the immense, erupting volcano that dominates the painting.[43]

The volcano of popular passions

Another aspect of what we might loosely term a Voltairean appreciation of the figure of the volcano is related to fanaticism, but encompasses a greater array of affects and attitudes. This is the 'volcano' of the human passions. Voltaire's protégé Sébastien-Roch Nicolas Chamfort produced a sustained meditation on this volcanic metaphor in his poem *Les Volcans, ode* of 1768. He begins by asking his muse to light up his 'genius' in order to paint the full horror of volcanic power. But he only does so the better to deplore more devastating human urges, notably the 'intrepid greed' of those exploiting the precious ores produced by volcanism, recklessly building their 'superb cities' on the still-fuming abyss. From there his censure turns to the terrible blood-lust of war and conquest, both past and present. The poet concludes that the ferocity of human emotions outstrips its volcanic model, so that 'man is forever the most destructive scourge of his fellow men'.[44] Interestingly, in the same year the playwright François-Thomas-Marie de Baculard d'Arnaud gave a Gothic twist to the same argument. Defending the exaggerated expression of the earthly passions on display in his convent drama *Euphémie*, the author argues that such intense emotions, confined and unnaturally stifled for too long, will burst out impetuously like the eruption of a volcano. The resulting conflagration 'is all the more terrible in that its fire has been compressed while being constantly stoked'.[45]

The English poet William Cowper was more political in identifying volcanic passions specifically with those inspiring over-reaching rulers.

43. Other artists were more respectful towards the Church. See, for example, the paintings by the Sicilian painter Alessandro d'Anna of the 1779 and 1794 eruptions of Vesuvius, which both feature stately, and seemingly effective, processions of a large icon of the Virgin Mary, preceded by orders of flagellants and penitents and surrounded by devout crowds of *lazzaroni*. Alessandro d'Anna, *Eruzione del Vesuvio dell'8 agosto 1779 con la processione dell'Immacolata* (1779), Acireale, Biblioteca e pinacoteca Zelantea; *Eruzione del Vesuvio del 1794, con la processione dell'Immacolata* (1794), Naples, private collection.
44. Sébastien-Roch Nicolas Chamfort, *Œuvres complètes de Chamfort*, 5 vols (Paris, 1824-1825), vol.5, p.126-29.
45. François-Thomas-Marie de Baculard d'Arnaud, *Euphémie, ou le Triomphe de la religion: drame*, 2nd edn (Paris, Le Jay, 1768), p.54.

His poem 'Heroism' (1782) uses the explosive figure of Etna to make his point:

> Behold in Ætna's emblematic fires
> The mischiefs your ambitious pride inspires.

This comparison then begs the rhetorical question:

> What are ye monarchs, laurel'd heroes, say,
> But Ætnas of the suffering world ye sway?[46]

In contrast, the heroism of the title refers to the stoic resilience of the peaceable country-folk, the 'swains' who deplore the depredations visited upon the land by their proud, warring overlords. After volcanic eruptions or the ravages of war, these commoners set patiently to recultivating their crops and resettling their flocks. The poem ends with a bathetic appeal to British patriotism, which somewhat dulls its radical political edge:

> Oh place me in some heav'n-protected isle,
> Where peace and equity and freedom smile,
> Where no Volcano pours his fiery flood,
> No crested warrior dips his plume in blood [...]
> A land that distant tyrants hate in vain,
> In Britain's isle, beneath a George's reign.[47]

These lines also suggest that Cowper subscribed to a more general climate theory holding that the absence of active volcanism in the British Isles corroborated, or even contributed to, its moderate form of government.

One of the chief theorists of eighteenth-century climate theory, Charles-Louis de Secondat de Montesquieu, had already drawn on images of natural catastrophes in order to explain the irresistible sweep of history. In his seminal *Considérations sur les causes de la grandeur des Romains et de leur décadence* (1734), he likened Tiberius' tyranny to a river bursting its banks and flooding the neighbouring lands with the unchecked abuses of imperial power.[48] However, in Montesquieu's classical historiography, the figure of subterranean volcanic fires

46. William Cowper, *Poems by William Cowper* (London, J. Johnson, 1782), p.357-62.
47. Cowper, *Poems*, p.361-62.
48. Charles-Louis de Secondat de Montesquieu, *Considérations sur les causes de la grandeur des Romains et de leur décadence* (Oxford, 2000), p.193.

was not applied to the overweening desires of kings and emperors, as it would be in Cowper's poem. Instead, it was reserved for the smouldering ambitions of the plebeian orders. 'While Rome was conquering the universe', Montesquieu writes, 'there was a hidden war within its walls: it was the kindling glow of fires, like those of volcanoes which burst forth as soon as some matter comes to fuel their combustion.'[49] The matter in this instance was the promise of greater civil liberties made to the plebeians if they would help the patricians to overthrow the monarchical order – which they duly did. With Montesquieu, the aristocratic Bordeaux magistrate who identified with the patrician Roman senate, the volcano thus begins to hint at the enormous dormant power of the masses.

As we have seen in Burke's *Enquiry*, and as will be manifest in responses to the French Revolution, this association of volcanic outpourings and popular uprisings was to prove influential and enduring. In a sense, from about 1750 onwards, it had an empirical as well as imaginative component. As increasing numbers of grand tourists experienced directly the fiery thrill of Vesuvius or Etna erupting, they also had impressed upon them the loud, tumultuous passions of the lowest orders of society in southern Italy. Many others enjoyed the spectacle of volcanic eruption remotely, in cheaper forms of reproduced tableaux and prints, in frequent newspaper reports and at touring diorama shows. These representations often consolidated in more oblique fashion the same association of the volcano and the mob through their condescending allusions to the intemperate, superstitious populace. As early as 1717 Bishop George Berkeley wrote to Alexander Pope rhapsodising about the volcanic island paradise of Ischia off the coast of Naples. It offered the visitor a lush 'romantic confusion' of scenery and flora which was, however, blighted by its inhabitants. The locals, wrote the bishop sniffily, 'have got […] an ill habit of murdering one another on slight Offences'.[50] As proof, he cites the incident of a youth shot dead on his doorstep two nights after his party had arrived on the island. By 1765, when the French astronomer Joseph Jérôme Lalande first arrived in Naples, casual asides about the hot-bloodedness of the inhabitants were no longer enough to satisfy

49. Montesquieu, *Considérations*, p.145.

50. Reverend George Berkeley, bishop of Cloyne, to Alexander Pope, Friday 22 October 1717, Naples. *Electronic Enlightenment scholarly edition of correspondence*, http://dx.doi.org.sheffield.idm.oclc.org/10.13051/ee:doc/popealOU0010445b1c (last accessed 12 December 2018).

the traveller's curiosity or his reason. In fact, Lalande devotes a whole chapter of volume six of his *Voyage d'un François en Italie* to the character and customs of contemporary Neapolitans. Overall, he paints a sympathetic portrait of a people whose most destitute and workshy orders have been 'humanized', as he puts it, by prolonged exposure to the cosmopolitan manners of the city's social elite.[51] He nonetheless signals the volubility, lasciviousness and shallow ostentatious piety typifying the *lazzaroni*, that is, the most chronically poor in Naples who numbered as many as 40,000 in a total population of approximately 400,000 by the end of the century.[52] Perhaps recalling the Roman sop to the plebs of 'bread and circuses', he notes that the Neapolitan urban poor are loud but ultimately docile and can be kept in order with three things: '*Farina*, *Furca*, *Festini*, food, pitchforks and festivals'.[53]

Lalande is typical of a grand-tourist set increasingly indulging what Melissa Calaresu calls their 'ethnographic impulse'.[54] The stereotypes of the *lazzaroni* as 'lazy, licentious and turbulent' are on occasion nuanced by later observers such as John Moore and Goethe or challenged by local writers like Giuseppe Maria Galanti.[55] Yet even in Galanti's guide to the city certain tropes persist, perpetuating the image of a street people, homeless and half-dressed, fond of song and cheerfully subsisting on macaroni and offal.[56] Hence, by the second half of the eighteenth century, they had become a picturesque urban complement to the picturesque volcano. Indeed, Richard de Saint-Non's influential *Voyage pittoresque, ou Description des royaume de Naples et de Sicile* (1781-1786) contains tableaux of Vesuvius erupting alongside street scenes in Naples, and even a historical reminder of the popular revolt against the Spanish Habsburgs led by the charismatic fisherman Masaniello in 1647.[57] If Saint-Non's pictorial

51. Lalande, *Voyage d'un François*, vol.6, p.336. More generally on early modern Naples, see *New approaches to Naples c.1500-c.1800: the power of place*, ed. Melissa Calaresu and Helen Hills (Abingdon, 2016).
52. Lalande, *Voyage d'un François*, vol.6, p.334-38, 343-44.
53. Lalande, *Voyage d'un François*, vol.6, p.335.
54. Melissa Calaresu, 'From the street to stereotype: urban space, travel and the picturesque in late eighteenth-century Naples', *Italian studies* 62:2 (autumn 2007), p.189-203 (192).
55. See Calaresu, 'From the street to stereotype', p.199-200.
56. Giuseppe Maria Galanti, *Breve descrizione della città di Napoli e del suo contorno* (Naples, presso li Soci del Gabinetto Letterario, 1792), p.267-68. Also cited in Calaresu, 'From the street to stereotype', p.200.
57. Saint-Non, *Voyage pittoresque*, vol.1, p.242, plate 103, 'Mazanielle haranguant

juxtaposition is suggestive of a 'volcanic' urban poor in Naples, a more concrete ethnographical link is established in the work of the artist Pietro Fabris. Only three years before he produced his stunning depictions of Vesuvius erupting for Sir William Hamilton's *Campi Phlegraei* (1776), Fabris had detailed the colourful crafts and fashions of the urban poor in Naples and its environs in his *Raccolta di varii vestimenti ed arti del regno di Napoli*.[58] A few of these picturesque figures reappear in miniature, studding the dazzling volcanic spectacles of his later work for Hamilton. Certain forms of local labour depicted by Fabris, such as quarrying, carry over directly from the illustrations of the *Raccolta* to those of the *Campi Phlegraei*. Noah Heringman goes so far as to suggest that the artist's precision in recording the geological strata of the mountain, delineated in different colours, echoes the ethnographic practice of the *Raccolta* in representing the varied and discrete social strata of the nearby city.[59] Thus a folkloric people colour contemporary perceptions of Vesuvius, just as the volcano's truculence informs accounts of the local population.

Such an identification of volcano and populace may be picturesque, but it is rarely benign. As hinted at in historical analogies and ethnographic portraits, the 'volcanic' lower orders were depicted as being ignorant, gullible and superstitious – qualities that sit ill with their dangerous volatility. This sense of benighted popular menace, a smouldering urban counterpart to the volcano, was given a veneer of scientific objectivity in Hamilton's influential accounts of the eruptions of Vesuvius in 1767 and 1779. In documenting the terrific outpourings of fire from the mountain, the British diplomat never fails to note the frenzied scenes of popular unrest that they trigger. In 1767 the eruption entailed a riot in the public jail and a 'mob' setting fire to the archbishop's gate because he had initially refused to release the relics of San Gennaro to the terrified populace.[60] Hamilton similarly mocked the popular recourse in Catania to the holy relics of Sant'Agata – which included her severed nipple – although it was her virgin's veil that was allegedly the most effective fetish for halting eruptions.

le peuple de Naples dans la place du marché des Carmes pendant la fameuse sédition de 1647'.

58. Pietro Fabris, *Raccolta di varii vestimenti ed arti del regno di Napoli* (Naples, n.n., 1773). On the relations between Hamilton and Fabris, see Heringman, *Sciences of antiquity*, p.95-109.

59. Heringman, *Sciences of antiquity*, p.102, 98.

60. Hamilton, *Observations on Mount Vesuvius*, p.32.

Inadvertently then, the ecclesiastical records of parading the saint's relics became useful markers of notable volcanic activity on Etna.[61] In 1779 Hamilton reports drily that the violent eruption of Vesuvius in the August of that year inevitably ignited 'the usual extravagant mix of riot and bigotry' among the ignorant masses clamouring through Naples. Yet contempt for the panic-stricken people often disguised a deeper-seated fear of them, especially among tourists. Some witnesses to the mutually reinforcing eruptions of volcanic violence and popular unrest openly professed a greater dread of the mutinous crowds than of the thundering volcano. Thus Hester Piozzi writes: 'Every time Vesuvius burns with danger to the city, they [the people] scruple not to insult their Sovereign as he flies from it; throwing large stones at his chariot, guards, etc.; making the insurrection it is sure to occasion, more perilous, if possible, than the volcano itself.'[62] In a century in which European penal codes punished public-order disturbances and threats to private property more severely than almost all other crimes, the volcano in full fiery spate was an apt symbol for fears of popular revolt, even revolution, among the educated and propertied elites. This symbolism was possibly overdetermined for those who experienced it *in situ* in the Bay of Naples.

The volcano as a source of enlightenment

One response to the fearful aversion felt by the educated classes towards the wretched and ignorant was to recast the volcano more as a source of light than as a seat of intense heat. Hamilton thus stresses the erupting volcano's brightness at night, explaining that 'the smallest objects cou'd be clearly distinguished, at any place within six miles or more' from the mountain when it exploded into life in August 1779.[63] But this should not induce the observer to conclude that volcanoes burn more brightly at night: that is 'vulgar' thinking. He expands on this implicit social distinction by contrasting the 'dishevell'd heads' of the Neapolitan hoi polloi with the calm study exhibited by one 'Mr Morris, an English Gentleman' at Sorrento who was able 'to read the title page of a book by that Volcanick light'.[64] His antitheses are intentionally stark. On the one hand, we have benighted popular

61. Hamilton, *Observations on Mount Vesuvius*, p.86.
62. Piozzi, *Observations and reflections*, p.321.
63. Hamilton, *Supplement to the 'Campi Phlegraei'*, p.8.
64. Hamilton, *Supplement to the 'Campi Phlegraei'*, p.12.

superstition and terror; on the other, the phlegmatic Englishman, the light, the learning, the enlightenment. In this way, Hamilton revives a cultured commonplace about reading at night by the light of a distant but luminously active volcano found in reports of earlier eruptions going back at least as far as Ellis Veryard in 1701.[65] But he also consecrates a trope that gazettes of the time would popularise in their accounts of the likes of Prince Borghese, finding himself at thirty miles from the eruption, yet easily able to read a letter by the same nocturnal volcanic glow.[66] In all these instances, however, volcanic enlightenment does not illuminate the lower reaches of society, plunged as they are in the perpetual gloom of ignorance, superstition and fear.

Other visitors preferred to follow Hamilton in his further characterisation of the local population as sempiternally indifferent to the volcano. These are the fishermen and street sellers on the quays who do not turn around to gape at the billowing columns of fire and ash, but continue with their daily toil, in the paintings of Bonavia, Volaire and others. Sir James Edward Smith was thus surprised to find that 'scarcely one among a hundred' of the city dwellers 'can be found who has been upon that mountain'.[67] Piozzi, though, received more nuanced impressions from a cynical Franciscan monk who remarks that the volcano 'throws up money' for the Neapolitans insofar as it attracts an endless trade in wealthy, gawking foreigners.[68] A canny peasant on the slopes of Vesuvius was similarly calculating. He told Piozzi that three years of uninterrupted cultivation in the ash-rich soils represented a small fortune for which it was worth risking possible rivers of fire and deadly volcanic hail.[69]

However, for enlightened observers of the volcanoes of southern Italy in the eighteenth century, there was another philosophical lesson to be drawn from the timeless incuriosity of the local population. If the volcano constituted in some sense their eternal lot, this was principally because the volcano itself was eternal. This assumption was both empirical and polemical. Empirically, it arose from considered observations made in the field, like those of Patrick Brydone in Sicily

65. Veryard, *An Account of divers choice remarks*, p.216-17.

66. See the *Gazette d'Amsterdam* of 31 August 1779, as cited in Anne-Marie Mercier-Faivre, 'Eruptions dans la presse: l'image du Vésuve dans la presse européenne francophone du XVIIIe siècle', in *Le Vésuve en éruption*, ed. E. Beck Saiello and D. Bertrand, p.83-98 (96).

67. Smith, *A Sketch of a tour on the continent*, vol.2, p.93.

68. Piozzi, *Observations and reflections*, p.305.

69. Piozzi, *Observations and reflections*, p.353.

in 1770. He was shown a well shaft in Catania sunk through several layers of distinct lava flows, each topped with thin soils. Basing his calculations on local evidence, he deduces: 'If it requires two thousand years or upwards, to form but a scanty soil on the surface of a lava, there must have been more than that space of time betwixt each of the eruptions that has formed these strata.'[70] Hence Brydone arrives at the conclusion that the lowest of the lavas must be at least 14,000 years old.[71] This supposition is given pictorial support in Hamilton's *Campi Phlegraei* in an illustration of lava flows pouring from a secondary vent of Vesuvius and forming a minor cone on the charred mountainside (Figure 5). The image appears to corroborate Hamilton's key empirical deduction in his *Observations on Mount Vesuvius, Mount Etna and other volcanos* (1772) that 'Mountains are produced by Volcanos, and not Volcanos by Mountains'.[72] Thus Fabris' *in situ* depiction of small-scale volcanic activity implies that the entire towering mass of Vesuvius must have been accumulated by the same process over unimaginably long historical periods.[73]

The polemical angle of such findings is that they are blatantly anti-clerical. Contemporary biblical exegesis held that the Earth was approximately 6000 years old, and that mankind and the world were the products of nearly synchronous acts of divine creation as related in Genesis. As paleontological and geological proofs continued to accrue which implied a much more ancient planet, some in the Church produced sophisticated arguments to demonstrate the contrary. The seventeenth-century Irish protestant archbishop James Ussher is the best known of these biblical chronologists. He claimed to have calculated the exact date of the Creation: the world emerged from primeval chaos, it seems, through the night of 22 October 4004 BCE. However preposterous this close dating might seem to us, it was the result of extensive scholarship in ancient religious history and was not dissimilar to estimates made by Ussher's celebrated contemporaries Johannes Kepler and Isaac Newton.[74] By the mid- to late eighteenth century, geological fieldwork was nonetheless scotching

70. Brydone, *A Tour through Sicily*, vol.1, p.130.
71. Brydone, *A Tour through Sicily*, vol.1, p.132.
72. Hamilton, *Observations on Mount Vesuvius*, p.52.
73. See also Rudwick, *The New science of geology*, ch.5: 'The emergence of a visual language for geological science', p.175. On both Hamilton and Brydone's 'eternalist' calculations of the age of volcanoes, and thus of the Earth, see Rudwick, *Bursting the limits of time*, p.119-22.
74. On Ussher, see Rudwick, *Bursting the limits of time*, p.115-18; for other views on

Figure 5: Pietro Fabris, 'Crater of Mount Vesuvius' (1776), in Sir William Hamilton, *Campi Phlegraei: observations on the volcanos of the Two Sicilies*, 2 vols (Naples, n.n., 1776), vol.2, plate 10. Private collection.

these abstract calculations and could no longer be ignored by the Church. Thus from the 1770s onwards, orthodox Christian thinkers, such as the scientific publisher Antonio Felix Castrioto in Portugal or the eminent earth scientist Jean-André de Luc in Britain, attempted to co-opt geological research into their demonstrations of the accuracy of biblical chronology. De Luc's *Observations sur la physique, sur la chimie, sur l'histoire naturelle et sur les arts*, published in a number of volumes in the 1790s, represent the most concerted and plausible reconciliation of the biblical text and nascent earth sciences.[75]

Having said that, the geological evidence attesting to vast swathes of prehistory before the advent of the human species, let alone revealed religion, was becoming overwhelming. This is patent not just in the likes of Brydone and Hamilton's fieldwork but also in the way in which their accounts often mocked Church authorities into the bargain. Brydone, for example, teases his host in Catania, the Catholic canon Giuseppe Recupero, with the glaring contradiction between the assertions of his faith and the findings of his geology. According to Brydone, Recupero himself admitted as much: 'Moses hangs like a dead weight upon him', writes Brydone, since 'he has not the conscience to make his mountain so young, as that prophet makes the world.'[76] In the case of Hamilton's image of volcanic orogenesis, the subversion of scriptural lore is more subtle, involving the engraving's usurpation of a 'quasi-aerial' divine vantage point from which to present the pre-biblical antiquity of the mountain.[77]

The construction of a deep-time geology was particularly grist to the mill of deists, such as James Hutton or Voltaire, who were keen to retain a belief in a Supreme Being behind the world's 'wise œconomy', as Hutton puts it,[78] but who also sought to undermine Christian orthodoxy in matters of Earth chronology. For Voltaire, it appeared 'quite simple that the eruptions of volcanoes produce small mountains'. However, he contrasts this relatively small-scale volcanic mountain building with what he sees as the eternal presence on the planet of

---

biblical chronologies among early geologists, especially the physico-theologists, see Rappaport, *When geologists were historians*, p.189-99.

75. On Castrioto, see Manuel S. Pinto and Filomena Amador, 'Discussing the age of the Earth in 1779 in Portugal', in *Geology and religion*, ed. M. Kölbl-Ebert, p.83-87; on de Luc, see Rudwick, *The New science of geology*, ch.6: 'Jean-André de Luc and nature's chronology'.
76. Brydone, *A Tour through Sicily*, vol.1, p.132.
77. Rudwick, 'The emergence of a visual language for geological science', p.175.
78. Hutton, *Theory of the Earth*, vol.1, p.223.

sweeping mountain ranges like the Alps, Pyrenees or Andes, as 'ancient as the world'.[79] Thus Voltaire is ultimately less interested in volcanoes per se than in their contribution to humbling humanity into conceding its belated, tiny and insignificant place in the cosmic design.

Deist thinkers then wanted to present volcanic activity as a more or less eternal feature of the Earth's history. They also drew on Buffon's 'actualism' to maintain that volcanism was an observable and operative phenomenon that acted in today's world just as it had consistently done from the distant dawn of geological time. In other words, it was a common occurrence, not an extraordinary, let alone supernatural, manifestation of divine will. The corollary of this assumption that the volcano belonged to an immanent natural order was that it should be considered as fundamentally 'good'. Seeing the volcano as a positive element of benevolent nature at work, deists preferred to emphasise its long-term consequences over its devastating short-term effects. Over time, the volcano is ultimately construed as an agent of natural renewal, creating new lands, reinvigorating exhausted soils, releasing a wealth of mineral and metal ores. The volcano is thus clearly 'one of nature's good deeds', as the *Encyclopédie* has it.[80] Hamilton is of this school of thought. In his *Observations on Mount Vesuvius, Mount Etna and other volcanos*, he pictures volcanic subterranean fires working 'as moles in a field, throwing up here and there a hillock', or as 'the great plough [...] which Nature makes use of to turn up the bowels of the earth, and afford us fresh fields to work upon'.[81] Elsewhere in the same work volcanoes are conceived of as giant telluric 'tumours' gathering and releasing the pent-up 'humours' of the Earth and so maintaining its rude health.[82] The analogies all imply that volcanism is natural, eternal, actual and salutary.

It struck Hamilton's waspish but perceptive correspondent Horace Walpole that the deist conception of volcanoes, shared by Hamilton and Voltaire, marked a departure from older, less benign and more

79. Voltaire, *Correspondence and related documents*, ed. Th. Besterman, in *The Complete works of Voltaire*, vol.85-135 (Oxford, 1968-1977), vol.124, p.26-27, Voltaire to Sir William Hamilton, 17 June 1773 (D18429).

80. D'Holbach, 'Volcans', p.443. It is estimated that volcanic soils cover 1 per cent of the Earth's surface but constitute 10 per cent of cultivated land, although this is not without complications. See Vincent E. Neall, 'Volcanic soils', in *Encyclopedia of land use, land cover and soil sciences*, vol.7, ed. Willy H. Verheye (Oxford, 2006), p.23-45 (23-24).

81. Hamilton, *Observations on Mount Vesuvius*, p.91, 161.

82. Hamilton, *Observations on Mount Vesuvius*, p.108-109.

mechanistic views on the subject. He writes: 'We folks of old-fashioned understandings look on burning Mountains as very petulant ovens, & a little destructive. The modern French Philosophers seem to have a mind to make them Parents of Order, & a kind of Providence, as far as they will admit any.'[83] Walpole goes on to season this critique with a humorous allusion to the naturalist George Montagu who stated that the last earthquake experienced in England 'was so tame one might have stroked it'.[84] The wit notwithstanding, Walpole touches here on the delicate issue that opened this chapter: the extent to which the volcano is an explicit proof of the deity's providential plan. He hints at a subtle, but significant, distinction between a *deist* interpretation of volcanoes and a *theist* interpretation of volcanoes. For the deist, the volcano is a secondary element in a much more extensive and inscrutable cosmic plan. For the theist, by contrast, volcanoes can be 'read' as a divine code inscribed in nature as a part of the discernible 'text' of God's design. The theist's worldview is every bit as revealed as the holy scriptures of Christian, Muslim or Jewish religions, but it is revealed immediately, directly and often sensually, by nature itself, without the need for prophetic intercession or deference to sacerdotal castes.

Thus in Jean-Jacques Rousseau's *Discours sur l'origine et les fondemens de l'inégalité parmi les hommes* (1755) volcanoes appear as the happy natural sources of mankind's acquisition and mastery of fire as well as the likely inspiration for smelting ores to provide the first workable metals.[85] They are providential in their example. Rousseau's disciple Jacques-Henri Bernardin de Saint-Pierre goes further still in integrating the volcano into his all-embracing concept of the 'harmonies' of nature that regulate the physical world for the good of humankind. Bernardin's geology is strongly Neptunian, centred on an idiosyncratic theory of the world-shaping power of tides.[86] According to his *Etudes de la nature* (1784) volcanoes owe their existence

83. Horace Walpole to Sir William Hamilton, 13 August 1773, in *Horace Walpole's correspondence*, ed. W. S. Lewis *et al.*, 48 vols (London and Oxford, 1937-1983), vol.35 (1973), p.413.

84. Walpole, *Horace Walpole's correspondence*, vol.35, p.413.

85. Rousseau, *Discours sur l'origine et les fondemens de l'inégalité parmi les hommes*, p.89, 98-99.

86. See Joël Castonguay-Bélanger, 'Le sort de Galilée: *Paul et Virginie* et la théorie des marées de Bernardin de Saint-Pierre', *Eighteenth-century fiction* 20 (2007-2008), p.177-96. On volcanoes in Bernardin's work, see Jean-Michel Racault, 'Géologie, vulcanologie et imaginaire chez Bernardin de Saint-Pierre', *Revue italienne d'études françaises* 1 (2011), p.37-54.

and activity to the seawaters that they ingest and vomit as though they were gigantic coastal filters, lit and fuelled by the 'oils, bitumen and nitrates' from animal and vegetable debris that they helpfully siphon up and burn off, thereby preventing this fatty flotsam from polluting and choking up the seas. Volcanoes refresh the oceans as thunder purifies the air, or as a 'gardener, at the end of autumn, burns the weeds in his garden'.[87] A volcano can act in various beneficial ways, depending on where it is situated. Thus Hekla in Iceland has its fires extinguished every spring by thawing snows, with the meltwaters flushing southwards to revitalise tepid southern seas.[88] The providential function of active volcanoes is yet more striking among the islands of the Asian Pacific. Here they mark out land masses, especially at night, and thus serve as nature's very own 'lighthouses'.[89] On a more symbolic level, in his summative *Harmonies de la nature* (1795), Bernardin lists erupting volcanoes among the forces of natural discord, operating to check and balance the forces of natural concord, and so contribute to the beneficent equilibrium maintained by the 'divine harmonies' of nature.[90]

In essence, we seem to have come full circle. The theist interpretation of volcanoes, expressed at the end of the century, rejoins the physico-theological vision of volcanoes produced in the early 1700s. Fénelon's conception of volcanoes as providential founts of fire that humanity subsequently learned to reproduce in order to keep warm, cook food and drive industry differs only from Rousseau and Bernardin's philosophy in the former's attribution of them to a nominally Christian God. However, it could be argued that this difference is crucial and non-negotiable. The Rousseauist understanding of volcanism, thus freed from Christian eschatologies, becomes the spiritual equivalent of Buffon's 'actualism', intimating an eternal and ongoing relationship between humanity and nature that only needs be 'read' properly in order to be renewed. As a result, it reinvigorates a longing for communion with the natural world that would be more fully developed in early-nineteenth-century European Romanticism but which would

87. Jacques-Henri Bernardin de Saint-Pierre, *Les Etudes de la nature*, 3 vols (Paris, Imprimerie de Monsieur, 1784), vol.1, p.303-308.
88. See Jacques-Henri Bernardin de Saint-Pierre, 'Préambule' to *Paul et Virginie* (Paris, 1989), p.63-64.
89. Bernardin de Saint-Pierre, *Etudes de la nature*, vol.2, p.618-19.
90. Jacques-Henri Bernardin de Saint-Pierre, *Harmonies de la nature*, ed. Louis Aimé-Martin, 3 vols (Paris, 1815), vol.3, p.17-18.

also subtend the evolutionary biology of Jean-Baptiste Lamarck or the geographical 'Kosmos' project of Alexander von Humboldt.[91]

Prometheus versus Empedocles

Fénelon's physico-theology, like Rousseau's later providentialism, draws on humanity's mastery of fire – first by flints, then by wood fires and industrial forges. 'Man knew', he writes, 'how to kindle it and how to ply it to all his practices.'[92] This is also a Promethean undertaking. As the myth has it, Prometheus, the rebellious son of a Titan, stole fire from Olympus for the benefit of mankind and was horribly punished by Zeus for the theft, chained to a mountain in the Caucasus where an eagle was sent to eat out his heart. However, since Prometheus was immortal, his flesh grew back overnight, only for the eagle to resume its feast day after day for all eternity. In his *Psychanalyse du feu*, Gaston Bachelard uses the figure of Prometheus to explain the ambivalent nature of humanity's relationship with fire. On the one hand, the Titan's theft represents the domestication of fire; on the other, its cruel consequences stand as a symbolic warning of fire's deadly power. What Bachelard then calls the 'Prometheus complex' is a perception of fire as both tame instrument and intense torture. It champions the subjugation of fire to human needs, as a means of emulating our 'fathers' in their knowledge of fire and in surpassing them in new knowledge about it and in finding new uses for it. But the complex also recognises fire as the object of one of our first fundamental social prohibitions: do not touch fire, do not play with fire.[93]

In contrast to the Prometheus complex, Bachelard posits an 'Empedocles complex', named after the fifth-century-BCE Greek philosopher who, as legend has it, threw himself into the burning crater of Etna. The Empedocles complex conceives of fire primarily as a source of fantasy (*rêverie*) and as a means of immolation, especially self-immolation. To succumb to the Empedocles complex and to dreamily contemplate fire is to see the cosmic connections between 'the hearth and the volcano', between the destiny of the

91. See, for example, Pietro Corsi, *The Age of Lamarck: evolutionary theories in France 1790-1830* (Berkeley, CA, 1988); Margarita Bowen, *Empiricism and geographical thought: from Francis Bacon to Alexander von Humboldt* (Cambridge, 2009), p.240-59.
92. Fénelon, *Demonstration de l'existence de Dieu*, p.43.
93. Bachelard, *La Psychanalyse du feu*, p.21-27.

log in the grate and the destiny of the entire world. Such a person 'hears the siren call of the pyre; for him, destruction is more than a change of state, it is renewal.'[94] The Empedocles complex therefore exerts an irresistible, albeit paradoxical, fascination: it combines love and death in the appeal of fire so that 'by one's sacrifice in the heart of the flames, that fleeting moment furnishes us with a lesson in eternity. This complete death leaves no trace behind it and ensures that we pass wholly into the beyond. It is losing everything in order to gain everything.'[95] Bachelard claims that this complex is particularly conducive to poetry. One late-eighteenth-century example might be André Chénier's beautiful lyric fragment depicting Heracles' self-sacrifice by fire on Mount Oeta, both poet and demi-god fully embracing what the psychoanalyst calls the 'lure of the pyre'.[96] Yet, as Bachelard recognises, the finest eighteenth-century expression of the Empedocles complex is Friedrich Hölderlin's haunting, unfinished verse-drama *Der Tod des Empedokles* of 1798-1799.[97]

Hölderlin wrote three aborted versions of his 'mourning-play' about Empedocles' suicidal leap into Etna. With each successive version, the poet affirms more strongly both the tragedy and the heroism of the philosopher's fatal act. In the third and final fragment of the play, Empedocles faces a series of trials which seek to deny or alter his fate. Firstly, his friend Pausanias wants to accompany the philosopher to his volcanic doom; secondly, his alter ego or 'evil spirit', the Egyptian priest Manes, mocks his pretentions to divine union with the elements of earth, air and fire; finally, a Greek chorus of Empedocles' fellow citizens from Acragas (Agrigentum), who have ostracised him from their city, attempt to call him back from the precipice.[98] As far as the text goes, Empedocles rejects each of them in turn and pursues his lone course to hurl himself into the volcano. His is a Christ-like individual self-sacrifice, surrendering in his person all human art to nature, but in doing so, however fleetingly, he gives meaningless nature meaning, endowing universal nature with a particular form. In Hölderlin's drama Etna becomes once again the site of the sublime, but in the sense of *sub*

94. Bachelard, *La Psychanalyse du feu*, p.31-40.
95. Bachelard, *La Psychanalyse du feu*, p.36.
96. André Chénier, 'La mort d'Hercule', in *Œuvres complètes* (Paris, 1950), p.27; Bachelard, *La Psychanalyse du feu*, p.40.
97. Bachelard, *La Psychanalyse du feu*, p.38-39.
98. Friedrich Hölderlin, *Der Tod des Empedokles*, in *Sämtliche Werke*, ed. Friedrich Beissner, 15 vols (Stuttgart, 1943-1985), vol.4:1 (1961). For the third version of the play, see p.119-41.

limen, forming a vertiginous 'threshold' from which the philosopher is poised to jump.[99] It is a threshold which both separates and unites opposites: snow and flame, earth and air, art and nature, the human and the divine. Only Empedocles' leap into the purifying fires of the volcano can bring these opposites together in an act of self-sacrifice that is also a moment of pure transcendence.[100] Like the blazing pyre of Chénier's 'Heracles', this is volcanic fire as a means of apotheosis – man becomes a god, but nature too is re-enchanted in the process.[101]

In a dense philosophical note written in preparation of the third iteration of his play, the so-called 'Grund zum Empedokles', Hölderlin explains that nature and art exist in a state of 'harmonious antagonism'. Nature is 'aorgic' (*aorgische*) – a neologism meaning inchoate, anarchic, excessive – while art is 'organisational', that is, crafted, structured, contained.[102] The volcano (nature) thus confronts the *polis* (art). These antagonistic contraries can only be fully realised in their union, but that union can only be realised in Empedocles' self-sacrifice in Etna. His mystical merger with the volcanic fires is coalescing as collision (harmony *qua* antagonism). The tragedy resides in the fact that only his death can bring about this moment of 'pure life'.[103] It is equally tragic, as Hölderlin explains, that even the 'all-knowing' Manes cannot comprehend what Empedocles experiences, since 'such a life is only accessible to feeling, and not to knowledge'.[104] In terms of the Empedocles complex, the eponymous philosopher goes to meet his 'free death' as a liberation and a completion, standing in stark contrast to Prometheus' fate – forever bound, forever incomplete.[105] Empedocles

99. On Hölderlin and the sublime, see Marc De Kesel, '"Wie der Gott und Mensch sich paart": Hölderlin's theory of the sublime as read in his "Remarks on tragedy"', in *Histories of the sublime*, ed. Bart Vandenabeele (Brussels, 2005), p.123-33.

100. This resembles the astrophysical notion of a 'singularity', such as the black hole. Empedocles' body, as it falls into Etna's crater, becomes a point of infinite density, exerting an irresistible pull on the elements around it: even light ('ether') and matter ('earth') cannot escape it, but are drawn into a fiery fusion with it, in a moment where the laws of physics no longer operate.

101. See Thüsen, *Schönheit und Schrecken der Vulkane*, 'Epilog: die Trauer des Empedokles', p.185-96.

102. Friedrich Hölderlin, 'Grund zum Empedokles', in *Sämtliche Werke*, vol.4:1, p.149-62 (152).

103. Hölderlin, 'Grund zum Empedokles', p.152-53.

104. Hölderlin, *Der Tod des Empedokles: dritte Fassung*, p.136, 139; 'Grund zum Empedokles', p.152.

105. Hölderlin, *Der Tod des Empedokles: dritte Fassung*, p.139.

'interpenetrates' fiery Etna, whereas Prometheus is chained coldly to the inert Caucasus.[106]

Clearly, Hölderlin is not interested in Etna as a physical volcanic topography, but as a dramatic 'zone of transcendence'.[107] He had never visited southern Italy, and most of what he knew of volcanic landscapes came from his friend and mentor Wilhelm Heinse who had both toured and fictionalised Vesuvius.[108] Nonetheless, if we scrutinise more down-to-earth accounts of confronting active volcanoes in eighteenth-century Europe, we still find clear and frequent traces of both the Empedocles complex and its Promethean counterpoint. Charles Pinot Duclos's swift dismissal of venturing onto the 'dangerous' slopes of Vesuvius, for instance, is redolent of a cultivated respect for the dangers of fire.[109] Goethe too on his first visit to the same volcano in March 1787 is cautious and turns back as soon as he feels unsteady on its treacherous lava screes. In later ascents of Vesuvius, the German poet similarly consults his guide to ensure that the terrain is safe.[110] Yet, if both Duclos and Goethe represent a certain culture of prudence regarding volcanic fire, the latter also expresses a Promethean desire to know and to control the volcano's blaze. Even before arriving in Naples, Goethe thrilled in advance at the prospect of taking possession of Vesuvius and its lava flows, eager 'to make these mighty objects my own'.[111] Once there and approaching the exploding, smoke-wreathed crater, he exults that 'there is something captivating about a present danger which challenges Man's spirit of contradiction to defy it.'[112] So this Promethean spirit advances onto a 'sharp edge of the monstrous abyss' or the vaulted roof of a purling lava flow,[113] but he always does

106. Hölderlin, 'Grund zum Empedokles', p.159. Prometheus is referenced in the second version of the play, as a warning of what happens if humanity is overexposed to the divine; see Hölderlin, *Der Tod des Empedokles: zweite Fassung*, p.92. See also the excellent analysis in the English translation by David Farrell Krell, in Friedrich Hölderlin, *The Death of Empedocles: a mourning-play* (New York, 2008), p.275-306, especially the notes on Prometheus and Empedocles, p.287, which suggests that they may have more in common than Bachelard's antithetical complexes suppose.

107. Thüsen, *Schönheit und Schrecken der Vulkane*, p.193.

108. Thüsen, *Schönheit und Schrecken der Vulkane*, p.186-87.

109. Duclos, *Voyage en Italie*, p.279.

110. Goethe, *Italienische Reise*, p.189, 195, 216.

111. Goethe, *Italienische Reise*, p.174.

112. Goethe, *Italienische Reise*, p.195.

113. Goethe, *Italienische Reise*, p.195, 216-17.

so with the intent of wresting some new object of knowledge from it. And if he does not steal fire itself from the volcano, he returns laden with new mineral-rich lava samples, not to mention the exciting tales of his incursions into 'the realm of Pluto'.[114]

In contrast, for visitors to Vesuvius such as Roland de La Platière or Louise-Elisabeth Vigée-Lebrun, the mesmerising fires of the volcano plunge the spectator into deep and dangerous reverie. Each deafening, dazzling eruption then becomes an invitation to take risks, without heed of friends or guides, an occasion more to transgress deep-rooted psychological taboos than prevailing cultural norms. Vigée-Lebrun acknowledges her fear of the volcano, amplified in the tearful terror expressed by her daughter whom she had brought to the burning spectacle. But her fascination for the volcano harbours a darker desire, a lover's identification with its flames: 'It would take little for me to make myself Vesuvian, I so love this superb volcano.'[115] This glimpse of a nascent Empedocles complex in the exiled French portraitist is more explicitly voiced in Roland's accounts of his reckless ascents of the erupting volcano. When he slips down a precipice and hovers a hundred feet above a glowing river of lava, his party quails at his risk taking. 'Seeing the danger I courted,' he writes, 'my companions turned pale and trembling.'[116] But Roland prefers to imagine himself alone on the volcano, in a cut-and-thrust duel with it, a duel which is just as much a passionate lovers' two-step. Significantly, it is at night, in his dreams, that his Empedoclean desire for an all-consuming union with the volcano emerges most insistently. The physical heat of the volcano fires the emotional heat of his passion for it. Consequently, he leaves Vesuvius, in his own words, 'with the regrets of a lover, leaving forever his darling mistress'.[117]

If Roland left the volcanic south of Italy with a heavy heart, he soon found distraction in the hot embrace of French revolutionary politics. Conversely, Marie-Antoinette's portraitist and intimate, Vigée-Lebrun, fled the outbreak of Revolution in 1789 and found in southern Italy a certain solace in the natural violence of the volcano

114. Goethe, *Italienische Reise*, p.217; the original is 'schon auf etwas Plutonisches hindeuten', p.193, but this translation from J. W. Goethe, *Italian journey*, translated by W. H. Auden and Elizabeth Mayer (London, 1962), p.193, captures the poet's meaning just as well.

115. Vigée-Lebrun, *Souvenirs*, vol.2, p.104.

116. Roland de La Platière, *Lettres écrites de Suisse*, vol.4, p.33.

117. Roland de La Platière, *Lettres écrites de Suisse*, vol.4, p.39, 51.

from the horrors of popular insurrection at home. In this sense, the Empedocles complex might be seen not just as the individual state of mind of those drawn irresistibly into volcanic fire when confronting it in person, but as the troubling psychological orientation of a whole generation. Thus Roland's account of his perilous encounters with Vesuvius was designed to appeal to his younger, sensitive fiancée Marie-Jeanne Phlipon, already a convert to radical republicanism before 1789.[118] Similarly, the eponymous hero of Antoine-Louis de Saint-Just's epic poem *Organt* is transported to Sicily by his guardian angel and sees:

> *Mount Etna*'s flames whirling
> And its flanks resonating, rumbling [...]
> He looked on these shuddering rocks,
> The burning entrails of Mount Etna;
> The majestic rubble where *Terror* dwells.[119]

He teeters a moment on the brink of the crater, captivated like Roland on Vesuvius by the volcano's shifting, scorching screes. When the poem was anonymously published in May 1789, its twenty-one-year-old author was unsuspectingly teetering on the brink of a dizzying revolutionary career that would take him to the heart of government and ultimately to the guillotine, as Robespierre's most fiercely loyal deputy.

Conservative critics of the French Revolution were just as quick to seize on the revolutionaries' delusive infatuation with 'volcanic' forms of power. They claimed that radical politicians were intoxicated by the hot fumes of talk about the 'general will' and the 'natural rights' of the people. It represented an unprecedented and dangerous idealisation of the ignorant, volatile masses. In this much, it amounted to an extreme political exacerbation of the instinctive fear felt by well-to-do travellers among the Neapolitan *lazzaroni*. For many conservative and moderate thinkers within France and outside it, this Empedoclean combination of recklessness and reverie was the prelude to terrible Europe-wide carnage and anarchy. As the disgraced French finance minister Charles Alexandre de Calonne wrote in exile in Coblenz, counter-revolutionary pressure must urgently be brought to bear on

118. See her memoirs written in prison in 1793: Marie-Jeanne Roland, *Mémoires de Madame Roland* (Paris, 1864).

119. Antoine-Louis de Saint-Just, *Organt*, in *Œuvres complètes* (Paris, 2004), p.119-300 (264, his italics).

France in order to prevent 'a volcano whose flames cannot be stopped from spreading into our sorrowful country and reducing it to ashes'.[120]

In this context, it is perhaps not surprising that the erstwhile theorist of the political sublime Edmund Burke expressed these anticipated terrors most vividly. In his *Reflections on the Revolution in France* (1790), he claims that the Jacobins in particular do not operate with a sufficient sense of fear of the popular forces they have released, but, from 'their sense of perfect safety', they are kept in sublime ignorance of the terrible reality of the revolutionary crowd by the abstracting power of their democratic ideals and egalitarian fancies.[121] According to Burke, two of the plausible models for this pernicious advancement of theory over practice are Georges-Louis Leclerc de Buffon and, significantly, Empedocles.[122] The linking of the two natural scientists might seem odd at first sight. But it may be that for Burke what they have in common is an abstract conception of the natural world structured around the four primordial elements of fire, air, earth and water. What is more certain is that Empedocles stands less for empirical observation of natural phenomena in Burke's text than for a suicidal desire to merge with their overwhelming might. The Greek philosopher is explicitly characterised in the *Reflections* as a troubled mystic who 'in cold blood, is said to have leapt into the flames of a volcanic revolution, *Ardentem frigidus Ætnam insiluit*'.[123] Thus a renewed Empedocles complex inspires a whole generation of French men and women to throw themselves into the red-hot heat of Revolution in a mad identification with its terrible power.

Burke's lesson is destined as much for his own countrymen as for his French correspondent. The warning to all is dire: one may jump enthusiastically into the volcano of Revolution, but the sublime experience one seeks to share is an illusion and will necessarily reveal itself to be truly terrible when viewed close-up. That is, the 'natural rights of man' are only flimsy metaphysical fantasies, not concrete political realities, while 'le peuple' is not at all virtuous, but fickle and

120. Charles Alexandre de Calonne, *De l'état de la France tel qu'il peut et qu'il doit être* (London, n.n., 1790), p.20. Cited in Mary Ashburn Miller, *A Natural history of revolution: violence and nature in the French revolutionary imagination 1789-1794* (Ithaca, NY, and London, 2011), p.146.

121. Edmund Burke, *Reflections on the Revolution in France* (Oxford, 1999), p.40.

122. Burke, *Reflections*, p.174.

123. Burke, *Reflections*, p.62. The citation comes from Horace, *Ars poetica*, lines 465-66: 'Empedocles [...] coolly leapt into burning Etna.' See Horace, *Satires, epistles, the art of poetry* (London, 1926), p.488-89.

barbaric. In this way the manly, patriarchal sublime once extolled by Burke, and incarnate in kingship in his *Philosophical enquiry into the origin of our ideas of the sublime and the beautiful* (1757), transmogrifies into the shrieking female, parricidal terror of the revolutionary *journées d'octobre* in 1789, as related in the *Reflections*.[124] In short, Burke rejects the volcanic model of political action. There is no need for a violent eruption to purge the earth and regenerate the land. Rather, a conservative evolution should be advocated which reforms gradually from generation to generation, building on what necessity, tradition and ancestry bequeath us and on what increasingly enlightened common sense sees fit to amend. As we shall see, the revolutionaries in France envisaged and employed the figure of the volcano altogether differently.

124. Burke, *Reflections*, p.71-72.

5. A volcanology of revolution 1789-1794

The volcano is a significant and recurrent element in the aesthetics of revolution in late-eighteenth-century Europe. As a key figure of the natural sublime, the erupting volcano was used especially by revolutionaries in France from 1789 onwards to designate great acts or forces deemed to be beyond human imagination. In essence, there were three sorts of volcanic sublime in revolutionary politics:

i. the volcano as a figure of the 'natural' overwhelming might of the Revolution itself;
ii. the volcano as an image of the indomitable revolutionary orator;
iii. the volcano as the embodiment of the irresistible collective power of the insurgent *peuple*.

In terms of the 'natural' energies released in the upheavals of revolution, the recourse to an aesthetics of the volcanic sublime modifies the fundamental meaning of the term 'revolution', moving from its astronomical sense of cyclical return to one of unprecedented historical and political rupture with what has gone before. The volcano is ideal in rendering this new sense of revolution insofar as it is perceived as *the* natural break in the very order of nature, as exemplary of nature's supreme and sublime power of violent self-regeneration. By analogy, the political Revolution of 1789 was presented by its proponents as operating a similar historical break in the very order of history, embodying the violent self-regeneration of the French nation. Temporally, volcanic eruption and the advent of the Revolution mark a before and an after; the new natural order, as the new political order, is initially recognised negatively, defined by what it has destroyed forever (a natural and political *ancien régime*, so to speak). Moreover, as with other sublime phenomena, which become terrible when experienced from too close a vantage point, this 'volcanic revolution' gains as a political spectacle with distance, like Vesuvius viewed from the pier in Naples or Etna seen from the Greek theatre at Taormina. Hence the French *prise de la Bastille* has echoes in English and German with untamable forces of nature as a sublime 'storming' or *Ansturm* of the fortress-prison.[1]

At the outbreak of the Revolution, similarly sublime images of lightning and electricity often translated for their French public a sense of sudden and violent change.[2] Yet lightning, tamed scientifically by Franklin and Nollet, does not possess the same sublime energy or the same order of potential devastation as the erupting volcano. What is more, for the deputies of the National Assembly who came to Paris in the Revolution from the provinces after the elections of September 1791, the volcano had the further figurative attribute of dangerous latency. Especially for the radical federalists known as the Girondins, the dormant yet rumbling volcano provided an apt metaphor for the uneasy peace in Europe which smouldered between revolutionary France and its increasingly hostile and vociferous enemies in neighbouring monarchies and principalities; an enmity often stoked by *émigré* nobles and priests exiled abroad. By January 1792 relations between France and its conservative neighbours had degenerated to the point that the Girondin deputy Maximin Isnard compared this phoney war to the 'menacing tranquillity of Etna'. All appears silent and calm yet, he continued, 'open up [the mountain] suddenly and you will find an abyss of fire, torrents of lava that are preparing the next eruptions.'[3] Isnard's truculent rhetoric was of a piece with the Girondins' sustained warmongering in the period, since they believed that armed conflict alone could silence once and for all the massing, murmuring forces of counter-Revolution hemming France in on all sides. Figuratively, the Girondins wanted to fire the revolutionary volcano into action and, with a singular, overwhelming eruption, cleanse the air of all reactionary conspiracy.

In April 1792 the warmongers in the National Assembly got their wish. A Girondin ministry pushed Louis XVI into declaring war

1. See Jean-Pierre Dubost, 'The sublime adverse and its Sadian reverse: Kant, Sade and Schiller', *Eighteenth-century studies* 28:1 (1994), p.83-93 (92).
2. Aurelio Principato, 'L'éloquence révolutionnaire: idéologie et légende', in *Histoire de la rhétorique dans l'Europe moderne*, ed. M. Fumaroli, p.1019-37 (1022).
3. Maximin Isnard, in the *Gazette nationale, ou le Moniteur universel* 6 (6 January 1792). Cited in Ashburn Miller, *A Natural history of revolution*, p.146-47. I am indebted to this fascinating study for number of important references and insights in this chapter.

on Austria, then on Prussia when it rallied to Austria's defence. The king had his own ulterior motives for going to war, secretly hoping for the victory of France's enemies and his subsequent restoration to lost privileges and powers. The disorganised French war effort quickly turned to debacle, so that by August 1792 Prussian troops were in striking distance of Paris. In a popular republican uprising, Louis XVI and his remaining Girondin ministers were overthrown on 10 August 1792 and a revolutionary republic was declared in the wake of the dual bloodbaths of the purge of counter-revolutionaries in the Parisian prisons (2-6 September 1792) and the decisive French military victory at Valmy (20 September 1792). The new republic elected a new National Assembly, the so-called Convention, in which the Girondins were often cast as moderate republicans against the increasingly dominant voices of the far-left Montagnards, many of whom were also influential members of the radical Jacobin Club in Paris. The designation 'Montagnard' derived from the disposition of the seats in the debating chamber of the National Assembly situated in the converted riding-stables of the Tuileries. The Montagnards (or mountain men) thus occupied the highest banks on the left of the Convention, jeering and sneering at their more inclusive, constitutionalist Girondin opponents seated opposite. Both factions were observed rather mutely by a majority of deputies belonging to neither camp, the so-called 'Plaine' (plain) or more contemptuously the 'Marais' (swamp or marsh), seated in the middle of the National Assembly. In the tumult of debate, the political geography of the hall was more than just an imaginatively named seating plan; it informed the very language of the new revolutionary political culture and the popular perception of its representatives.

The first task of the Convention was to try the 'traitor' king, a process which crystallised the antagonisms between Girondins and Montagnards. Broadly speaking, the Girondins were federalist and constitutionalist; they questioned whether the Convention had the constitutional power to try the king in the first place and sought instead to put the question of Louis XVI's fate to the French people as a nation. The Montagnards were Jacobin centralisers, mostly elected in and around Paris, and claimed their right to try the king not from any constitutional authority but as the elected representatives of popular sovereignty, playing specifically to their power base among the clamouring Parisian sans-culottes. In this charged political environment, the figure of the 'volcano' of Revolution proved a useful complementary image to that of the 'Montagne'. Now it was the Girondins themselves who were to be consumed in the lava of red-hot indignation that poured from the erupting 'Mountain'. The Montagnards claimed to speak with the thunderous voice of 'natural' justice that called for the monstrous traitor-king to be tried and executed if necessary, overwhelming in its fiery righteousness any opposition, denounced as royalist compromise. Louis XVI was thus eventually tried by the Convention, found guilty and sentenced to death by a slim majority. He was guillotined on 21 January 1793. Yet the king's death did nothing to abate the hostility between Girondins and Montagnards, not least because it redoubled the forces of the coalition of monarchist European states leagued against republican France. As the political infighting became more viciously personal in the course of 1793, supporters of the Montagnards were wont to confound the Girondins with the vacillating, opportunistic deputies of the 'Swamp' or 'Marsh'. As opposed to the salubrious air and moral elevation afforded by the 'Mountain', the 'Swamp' was a miasma of petty interests and poisonous opinion, thus an appropriate lair for the so-called 'hydra' of Girondin federalism. As Mary Ashburn Miller has convincingly shown, this stark opposition of political topographies draws not just on the steeply banked seating in the Convention but on decades of enlightened discourses advocating the draining of fetid, disease-ridden marshes across France, translated into legislation by the National Assembly in December 1790.[4] The culmination of this battle between the Gironde and the Montagne, between the revolutionary lowlands and mountain peaks, came with the militia-backed, populist purge of the Girondin leaders from the Convention on 2 June 1793. This moment is celebrated in Pierre-Michel Alix's coloured print *Le Triomphe de la montagne* (1793), which depicts a band of joyous revolutionaries dancing around a liberty tree on the lighter uplands of a mountain while, from the summit, flames spout and lightning bolts dart lethally at vile swamp-monsters and drowning priests in a murky pool below.[5] The volcano had finally cleansed the marsh.

The Jacobin Club launched the *Journal de la Montagne* in June 1793, making explicit the links between the radicals in the Club and those grouped on the high benches on the left of the Convention. The same month the National Assembly briefly debated, then adopted, the new revolutionary constitution, often called the 'Constitution

4. Ashburn Miller, *A Natural history of revolution*, p.123-27.

5. Pierre-Michel Alix, *Le Triomphe de la montagne* (1793), Paris, Bibliothèque nationale de France, De Vinck, 5012.

montagnarde'. This radical bill of rights extended the social scope of the 'Déclaration des droits de l'homme et du citoyen' and consecrated the notion of popular sovereignty over that of national sovereignty.[6] It was put to a nationwide referendum on 9 August 1793 and, despite massive abstentions, was overwhelmingly approved. Hundreds, if not thousands, of addresses flooded into the Convention from local administrators, mayors, leaders of popular assemblies, declaring general joy at the new constitution and swearing to shed their last drop of blood in defence of the revolutionary republic. Some, such as the 'popular society of true sans-culottes and republicans' from the Checy district of Orléans, acclaimed not only 'the one and indivisible republic' but also 'the deputies of the sublime Mountain, which has given birth [...] to this masterpiece of nature'.[7] The constitution was thus the work no longer of men but of nature itself, conceived on the sublime heights of the revolutionary mountain. The following day, 10 August 1793, was the first anniversary of the bloody overthrow of the French monarchy. In celebration of this momentous occasion, one engraver turned the figure of the sublime mountain bringing forth the new constitution into a volcanic event. Louis-Jean Allais's *La Constitution républicaine* represents a jagged mountain summit exploding with bright radiance, emitting lightning bolts and grey-brown clouds of smoke. At its white-hot centre the words 'Droits de l'homme, Constitution de 1793' appear as though engraved on a double tablet of stone. The caption makes the biblical reference clearer still: 'The republican Constitution, similar to the stone tablets of Moses, issues from the heart of the Mountain amid lightning'.[8] The great biblical lawgiver is secularised, divine law emanates no longer from God above but from nature itself, and more specifically from within the revolutionary mountain. By making Mount Sinai into a volcano, Allais suggests too that revolutionary rights are as much about rupture, explosion and heat as they are about light.

As the sacred revolutionary law descended from the mountain, it was transmitted and interpreted by particular spokesmen, a political priesthood striving and inevitably failing to incarnate Rousseau's

6. See Marcel Gauchet, *La Révolution des droits de l'homme* (Paris, 1989), p.209-26.
7. *Archives parlementaires de 1789-1860: recueil complet des débats législatifs et politiques des chambres françaises*, ed. J. Madival, E. Laurent *et al.*, 82 vols (Paris, 1862-1914), vol.70, p.501.
8. Louis-Jean Allais, *La Constitution républicaine* (1793), Paris, Bibliothèque nationale de France, De Vinck, 6170.

sublime Legislator from his *Contrat social*.[9] It is here, as Aurelio Principato has argued, that the volcano of Revolution is embodied by successive 'volcanic' revolutionary orators.[10] French writers from the Romantic generation of the early nineteenth century, such as Nodier, Hugo, Lamartine and Michelet, consecrated this identification of the individual voice with the booming thunder of Revolution. But revolutionary oratory was already considered by many as sublime in 1789. In part this was because it made consistent claims to break with the laboured, formulaic models of eloquence associated with the *Ancien Régime* (even though revolutionary orators respect the conventions of classical rhetoric much more than they would care to admit). It was sublime because it gave vent to 'natural' aspirations to freedom, equality and social justice. It was also deafeningly loud, echoing the irresistible voices of the mountain cataract, the stormy sea, the howling gale or the erupting volcano. *Tonner* (to thunder) was at the heart of its capacity to *étonner* (to astonish). Thus stentorian voices in particular predisposed their revolutionary audience to feelings of sublimity. When this was coupled with a large imposing physique equally apt to arouse a sense of the sublime, and with an ugliness which Burke had claimed was 'consistent enough with an idea of the sublime' if 'united with such qualities as excite a strong terror', then the orator had every chance of moving and enthusing his audience.[11]

This was evidently the case with the two most famous 'volcanoes' of the Revolution: Honoré Gabriel Riqueti de Mirabeau and Georges-Jacques Danton. Their Herculean stature set off faces disfigured in each case by the ravages of smallpox, leaving the skin itself like molten lava.[12] Such was the ugliness of Danton, his top lip cleft and his nose split by a kick from a rampant bull in his youth, that Jacques-Louis David allegedly found him unpaintable (the true mark of the unrepresentable sublime). The revolutionary's portrait was left to a student of David who, according to François-Alphonse Aulard, painted him in close-up, 'digging out one at a time the marks of smallpox, the

9. See Jean-Jacques Rousseau, *Du contrat social, ou Principes du droit politique*, in *Collection complète des œuvres*, vol.1, p.187-360 (232-37).
10. Principato, 'L'éloquence révolutionnaire', p.1031.
11. Burke, *A Philosophical enquiry*, p.119.
12. Smallpox or *la petite vérole* had a more subtle connotation of volcanoes for the French, insofar as they commonly believed it originated from Naples and was thus known as *le mal napolitain*. See Jean-Christophe Abramovici, 'Sade et l'irruption du désir', in *Le Vésuve en éruption*, ed. E. Beck Saiello and D. Bertrand, p.171-77 (172-73).

crevasses, mountains and valleys of that havoc-wrought visage'.[13] Danton thus emerges as a man-mountain, or man-volcano, worthy of Jules Michelet's description of 'the sublime ugliness of a face [...] scrambled by small-pox, and which resembled the scoria of Vesuvius or Etna'.[14]

Both Mirabeau and Danton appear to have fostered the image of the volcano, not just in their eruptive outpouring of thunderous discourse, but also in an indomitable immobility adopted in the midst of the agitated clamour set off by their words. In this much, both orators relied on the sheer force and presence of their personality. As Aulard puts it, 'the true inspiration for Mirabeau is his ego. He is always great, simple, sincere when he speaks of himself in self-defence or self-praise.'[15] Danton too spoke about himself, bragged about himself, mocked himself at the tribune, yet he also singled out his opponents as 'the person known as Dumouriez [...] the individual called Marat', just as the Revolutionary Tribunal and guillotine were to individualise their victims.[16] What distinguishes these two volcanoes of Revolution from one another, however, is less their physical demeanour than their rhetoric. Mirabeau's speeches frequently resort to the classical figures of the sublime, as in his famous denunciation of impending national bankruptcy of 26 September 1789 in which the nation's dire financial predicament is painted successively as a 'terrible explosion', as the 'incalculable evils that this catastrophe will spew over France', 'these convulsions of despair and wretchedness', 'the universal conflagration' and 'the universal ruin' of 'hideous bankruptcy'.[17] Nonetheless, his volcanic imagery is deployed in compliance with the conventions of a prepared oration. Danton's oratory is less sublime in content than in form. It strikes its audience with its disregard for rhetorical niceties, often wholly improvised, without exordium or peroration, being a sort of discursive eruption, going straight to the point, in a syntax which is often disjointed, its parts broken or disarticulated – precisely the formal features of the rhetorical sublime as first defined by Longinus.[18]

13. François-Alphonse Aulard, *Les Orateurs de la Législative et de la Convention*, 2 vols (Paris, 1885-1886), vol.2, p.221.
14. Jules Michelet, *Histoire de la Révolution française*, 2 vols (Paris, 1998), vol.1, p.811.
15. François-Alphonse Aulard, *Les Orateurs de l'Assemblée constituante* (Paris, 1882), p.124.
16. See Aulard, *Les Orateurs de la Législative et de la Convention*, vol.2, p.212.
17. Madival, Laurent *et al.*, *Archives parlementaires*, vol.9, p.195-96.
18. See Mats Malm, 'On the technique of the sublime', *Comparative literature* 52:1 (winter 2000), p.1-10.

Nonetheless, 'volcanic' orators such as Mirabeau and Danton ultimately drew their revolutionary authority from the sovereign people whom they allegedly represented. But *le peuple*, too, already had its own set of volcanic attributes. Firstly, as we have already seen in travellers' accounts of Naples, eighteenth-century identifications of the erupting volcano with an unruly populace were commonplace, even if they were often condescending, when not openly disparaging. Yet this condescension betrays a latent fear of uncontrollable popular power. To some extent, this fear still haunts the French revolutionaries even as they extol the sublimity of the people-volcano and draw their own political legitimacy from it. This is most evident in the fraught politics of representation in the Revolution: both representation as a political delegation of sovereign power to the people's deputies, and representation as the language or images in which this delegation is couched. It comes down to a question of the exercise of popular sovereignty which resides inalienably in the people, and in the people alone. The volcano, as with other sublime figures, refuses to delegate or otherwise divide this conception of popular sovereignty, being at once unmediated and unrepresentable in all its awful power. So how can the people-volcano be sovereign and free in a political system grounded in representation, even if mandated by and from the same *peuple* (understood as age-restricted universal male suffrage in the 1793 constitution)? In the Revolution this broadly means: the people are sublime or the revolutionary orator is sublime, but they cannot both possess revolutionary sublimity at the same time without one politically compromising the other. Danton appeared to understand this with the institution of the Revolutionary Tribunal in March 1793, famously instructing the members of the Convention, who would vote for the Tribunal and possibly sit on it, to 'be terrible in order to spare the people from being so'.[19] That is, the people had often shown themselves to be the agency of revolutionary terror, especially in murderous popular uprisings, which in turn had inspired deputies such as Danton to sublime rhetoric in their defence and praise. Now the process was to be reversed by making the Tribunal an instrument of universal fear in order that the people might bask in the sublime feelings occasioned by the implacable justice that the court was to deliver in their name. In fact, unwittingly or not, the Convention's establishment of the Revolutionary Tribunal would fulfil this function only too well, providing a principal apparatus of the future reign of Terror.

19. Madival, Laurent *et al.*, *Archives parlementaires*, vol.60, p.63.

Staging the volcano of revolution

Before we look at the fateful confluence of the revolutionary Terror and the figure of the volcano, it is worth considering another form of popular representation in revolutionary France: the theatre. The theatre in the Revolution was largely a place of distraction from more troubling political realities, yet the opportunity for ideological inculcation afforded by the stage was irresistible to the revolutionaries, for whom the oratorical and dramatic arts seemed almost interchangeable. After all, they would argue, had not Cicero learned from the actor Roscius?[20] Yet the new politics required a new theatre. Talma, the greatest actor of the revolutionary period, thus rejected the stilted, classical idea of the 'beautiful', choosing to replace its 'sad chants' with the 'plebeian' model of those popular orators 'who astonished by the sublime traits of their unrehearsed eloquence'.[21] In this way the politically performative speech-acts of a Danton or a Saint-Just were to be translated into theatrical performances for the edification of their audiences. Consequently, as at the tribune of the Convention, the patriotic rhetoric of the revolutionary theatre frequently transgressed the clearly defined generic boundaries of the dramatic art – tragedy slipping into comedy, the epic into the sentimental, reducing the political speech-act to a pantomimic parody of itself. It was the particular genius of the radical atheist republican Sylvain Maréchal to take full advantage of this confusion of dramatic genres in his agitprop play *Le Jugement dernier des rois* (1793), in which he puts exalted tribune oratory in the mouths of his sans-culottes while involving his pantomime monarchs and the pope in degrading boulevard farce. As is widely known, this play also contains the most famous representation of a staged volcanic eruption in the Revolution, one designed to engulf in its purging, republican lava all of the crowned heads of Europe.[22]

The plot of the play concerns an old man deported to a deserted volcanic island by a debauched French king (more Louis XV than

20. See Peter France, 'Speakers and audience: the first days of the Convention', in *Language and rhetoric of the Revolution*, ed. John Renwick (Edinburgh, 1990), p.50-74 (72).
21. Cited in Aurelio Principato, 'Comment restituer l'action oratoire de la Révolution?', in *Une Expérience rhétorique: l'éloquence de la Révolution*, ed. Eric Négrel and Jean-Paul Sermain (Oxford, 2002), p.19-33 (32).
22. Sylvain Maréchal, *Le Jugement dernier des rois*, in *Théâtre du XVIIIe siècle*, ed. Jacques Truchet, 2 vols (Paris, 1972-1974), vol.2, p.1307-25.

Louis XVI) who is rediscovered in a not-too-distant future by a group of sans-culottes assembled from all the countries of Europe who have come to maroon on the island the last representatives of their vanquished monarchies and papacy. The old man, a convert by bitter personal experience to 'natural' republicanism, is hailed and saved by the sans-culottes as one of their own while the deposed kings, queen (Catherine II of Russia) and pope are left to squabble and scrap on the island before its volcano explodes and consumes them in red-hot lava. The utopian vision of deposed, deported and destroyed monarchy was a recurrent trope of Maréchal's pre-revolutionary and early revolutionary writings.[23] Yet the theatrical masterstroke of *Le Jugement dernier des rois* was the addition of an active volcano to the desert island hosting the fallen monarchs. As well as conveying the more general political valorisation of the Montagne in 1793, the republican volcano in Maréchal's play had more specific resonance at the time of its first staging in mid-October of that year. Its first public 'eruption' came a day after Marie-Antoinette's execution and at exactly the same time as the royal tombs at Saint-Denis were being desecrated by anti-monarchical extremists.[24] Maréchal's volcano thus consecrated and completed the regicide work of the Revolution and sanctioned the vandalism of royal and religious sites that would characterise the dechristianisation campaigns of the first phase of the Terror. It does so by presenting revolutionary history as both performative and catastrophist. The old man is thus told by one of the sans-culottes: 'The French people rose up. They said: *I don't want any more kings*; and the throne disappeared. They then said: *I want the republic*, and here we are all republicans.'[25] Here the play, as a form of theatrical representation, cleverly sidesteps the issues of political representation that dogged the revolutionary regime during the trial of Louis XVI and thereafter. For *Le Jugement dernier des rois* never enacts more than the moment of founding violence of the universal republic: its final scene is the volcanic death of all kings wherein 'nature' consecrates the republican usurpation of all monarchical power. Its theatrical representation is thus a performance of the very suspension of all political representation, staging anew each

23. See Maurice Dommanget, *Sylvain Maréchal: l'égalitaire, 'l'homme sans Dieu', 1750-1803 – vie et œuvre de l'auteur du Manifeste des égaux* (Paris, 1950), p.143-47, 265-67.
24. See the interesting analysis in Sanja Perović, *The Calendar in Revolutionary France: perceptions of time in literature, culture, politics* (Cambridge, 2012), p.127-50.
25. Maréchal, *Le Jugement dernier des rois*, p.1313. His italics.

evening the arrogation of sovereignty – that exceptional moment when the laying down of the law is still outside the law – while never going beyond it to explore the conserving violence needed to consolidate and perpetuate this power. Hence the play would not fall foul of later revolutionary theatre legislation which attempted to rid the stage completely of any representation of, or reference to, the monarchy.[26] It is clear too that the propaganda value of *Le Jugement dernier des rois* was not lost on the Committee of Public Safety, the principal executive body of revolutionary government in this period. They diverted precious supplies of saltpetre and gunpowder in a time of war to the Théâtre de la République so that the climactic volcanic eruption could be staged night after night.[27] Moreover, the play had a print run of around 20,000 copies, 6000 of which were bought up by the revolutionary authorities to be circulated among front-line troops. The play was also performed in major cities across France to a potential audience of 100,000 spectators.[28]

The significance of Maréchal's play as a tool of revolutionary propaganda lay in the way it mythologised the forces of Revolution. The volcano becomes nature's sans-culotte – good, free, republican and regicide, atheist and dechristianising; it realises the myth that it is ultimately nature's decree and not the vote of the Convention which put Louis XVI to death, that a Jacobin republic is the sole form of 'natural' polity and so must be extended, by force of arms if necessary, to all the nations of the globe currently languishing under unnatural, 'monstrous' royal rule. There is a double shift in meaning taking place here: in *Le Jugement dernier des rois* the volcano goes from being a privileged metaphor of Revolution to being its principal agent; secondly, it does so from outside of historical time, regenerating France not just in immanent political terms as the physical instrument of violent regime change, but also secularising the apocalyptic Last Judgement by fire of the play's title in order to give a transcendental dimension to its actions.[29] Conversely, the revolutionary volcano becomes an 'infernal' power only for the kings, queen

26. See Martin Nadeau, 'La politique culturelle de l'An II: les infortunes de la propagande révolutionnaire au théâtre', *Annales historiques de la Révolution française* 327 (January-March 2002), p.57-74.
27. See Dommanget, *Sylvain Maréchal*, p.262-63; also Daniel Hamiche, *Le Théâtre et la Révolution* (Paris, 1973), p.174-75.
28. Hamiche, *Le Théâtre et la Révolution*, p.174; also Perović, *The Calendar in Revolutionary France*, p.142.
29. See Perović, *The Calendar in Revolutionary France*, p.127-28.

and pope still in thrall to the superstitious ways of the *Ancien Régime*, for whom it will ultimately prove a living 'hell' on earth. Of course, Maréchal harnesses the natural sublime too in his volcanic mythologising of Jacobin Revolution. Thus those diverse forces able to embody a natural sublime on the desert island – wild beasts, a primitive tribe of sun-worshippers and the volcano – all respect the marooned old man as the incarnation of revolutionary political sublimity. He thus enjoys a sort of Kantian intimation of the sublime character of his own moral and political worth in the face of these dreadful figures of raw natural power. *Le Jugement dernier des rois* goes further in also reversing an element of the Burkean political sublime. For Burke, we might recall, the Revolution degraded the sublime power of the institution of the monarchy, presenting it to the world as a 'monstrous tragi-comic scene'. In contrast, Maréchal's play represents the monarchs of Europe as 'crowned monsters' who denature the political sublimity of the Revolution and must therefore be eradicated.[30]

Yet in order to make a sublime spectacle out of the unnatural ferocity of kings and their terrible punishment in the flames of the volcano, a safe aestheticising distance needs to be interposed between the public and the play. In *Le Jugement dernier des rois* this safe distance is guaranteed by the theatrical device of the play within the play: that is, the audience watches the brave sans-culottes from all over Europe abandon their monarchs and the pope to the volcano, then these good republicans retire to a ring of waiting, armed frigates to watch the sublime demise of their royal and papal enemies. The sans-culottes watching from a secure remove (whether on stage or in the wings) sanction the safety of all virtuous revolutionaries in the audience, since, however close the latter might physically be to the stage, neither the blood-thirsty monarchs nor the terrible violence of the erupting volcano can harm them. In short, the Revolution knows its own and protects them so that, secure in their righteous republicanism, the explosive finale is pure sublime for them. (Despite not having clear stage instructions to this effect, it would make ideological and dramatic sense that the families of 'savages', who visit the island to worship the sun and who fraternise with the sans-culottes as primitive fellow republicans, also leave the feuding monarchs and watch from the safety of their imagined pirogues.)[31]

30. Burke, *Reflections*, p.10; Maréchal, *Le Jugement dernier des rois*, p.1321.
31. I differ on this point from Ashburn Miller (*A Natural history of revolution*, p.159, 163) who claims the 'sauvages' are caught up in the eruption, either as an

The sans-culottes of Maréchal's play are, then, exemplars of the revolutionary people in action, just as the active volcano represents the irresistible might of the Revolution itself. According to Sanja Perović, the playwright first made the connection between volcanic eruption and historical change in the commentaries he provided for the catalogue *Antiquités d'Herculanum* (1781), which recorded the archaeological finds made at the site on the slopes of Vesuvius.[32] The violent eruption of August 1779 was still fresh in many people's minds, falling as it had done exactly 1700 years from the devastation of Herculaneum and Pompeii. In his research for this work Maréchal became particularly interested in the ancient dual cults of the sun and the volcano that he discerned in primitive rites practised around Vesuvius. His investigations touched on the Ausones, an ancient tribe living and worshipping in the shadow of the volcano – a reference that would explicitly recur in his later work *Voyages de Pythagore en Egypte* (1798-1799) where he notes:

> The Ausones, who had already adopted the sun as God and king, believed that they saw in the flaming volcano the wrath of that same power which had hitherto only manifested itself to them in the form of beneficence. In order to appease it, they condemned any ambitious person, hostile to their primitive equality, to be an expiatory victim: they threw him into the burning crater; and thereafter they would repeat this terrible punishment at each eruption and for as long as there were rash mortals willing to call themselves giants among their equals.[33]

In a sense, *Le Jugement dernier des rois* gives a revolutionary updating of this ancient sacrificial rite, punishing contemporaries full of 'ambition' or those self-styled 'giants', such as monarchs and the pope, who sought to subvert the natural order of primitive equality. The play also politicises this practice, since the human sacrifice made here to the volcano is not the result of a superstitious fear taking hold of the population in the wake of an eruption; it is the product of enlightened republicanism actively conspiring with nature to rid the world of both monarchy and papacy.

---

unintended oversight of Maréchal's dramaturgy or as the necessary 'collateral damage' of often indiscriminate volcanic revolution.

32. Perović, *The Calendar in Revolutionary France*, p.82.

33. Sylvain Maréchal, *Voyages de Pythagore en Egypte, dans la Chaldée, dans l'Inde, en Crète, à Sparte, en Sicile, à Rome, à Carthage, à Marseille et dans les Gaules, suivis de ses lois politiques et morales*, 6 vols (Paris, Deterville, 1798-1799), vol.5, p.38-39. See also Dommanget, *Sylvain Maréchal*, p.266-67.

It is also possible that the inspiration for the volcano in Maréchal's play was of more recent date. The striking prints of Alix and Allais that we have already mentioned, depicting the volcanic victories of the revolutionary Mountain in the summer of 1793, may have influenced the playwright's choice of explosive denouement. Maurice Dommanget adds a further possible pictorial source to this list, an anonymous engraving entitled *Sic transit gloria mundi* which portrayed all the symbols of religious and royal authority, crowns, mitres, sword, orb and sceptre, being hurled into the void by an erupting volcano.[34] This image in fact dates from before the Revolution and derives from the Renaissance emblematic tradition representing the 'bonfire of all human vanities'. Nonetheless, the volcano is central here to the destruction of the insignia of kingship and ecclesiastic power, and recalls a similar bonfire set up on the Champ de Mars in celebration of the third anniversary of the storming of the Bastille in July 1792. This blazing auto-da-fé of all the attributes of noble privilege was rendered as a sort of volcanic pyre by Jean-Louis Prieur in his *Tableaux historiques de la Révolution française* (1791-1795).[35]

We might even speculate that the highly erudite Maréchal had another, more exotic model for his 'small uninhabited but inhabitable island' that was 'half-volcanised'.[36] For an odd literary-historical parallel links the play's fictitious island with the eighteenth-century French colony of Ile Bourbon, today's Ile de la Réunion. The parallel consists in this: Ile Bourbon was, and still is, a volcanic island in the Indian Ocean, being literally 'half-volcanised' in the sense that of its two volcanoes, only one is historically active. In the eighteenth century this volcano, Piton de la Fournaise, had noticeably leapt into fiery life in 1774, in 1776 and again in 1791.[37] When the island was first claimed for the French crown in 1640 it was a deserted watering stop on the way to India, yet its first inhabitants were not French merchants or sailors but a dozen or so mutineers deported and marooned there for their part in an uprising at Fort Dauphin in Madagascar. On 19 March 1793 the Convention decreed that the colony's despised royal

34. Dommanget, *Sylvain Maréchal*, p.266.
35. See Rolf Reichardt and Hubertus Kohle, *Visualizing the Revolution: politics and the pictorial arts in late eighteenth-century France* (London, 2008), p.60-62.
36. Maréchal, *Le Jugement dernier des rois*, p.1307, 1310.
37. Jean-François Lénat, 'A brief history of the observation of the central area of Piton de la Fournaise', in *Active volcanoes of the southwest Indian Ocean: Piton de la Fournaise and Karthala*, ed. Patrick Bachèlery *et al.* (Heidelberg, 2016), p.3-21 (3-11).

name be abolished, to be replaced with its current title of Ile de la Réunion, symbolically chosen to commemorate the fraternal joining of forces of the provincial militia and the Parisian sans-culottes in the violent overthrow of the discredited constitutional monarchy on 10 August 1792.[38] Thus in *Le Jugement dernier des rois*, sans-culotte Réunion vanquishes forever monarchical Bourbon in keeping with the Convention's decree to rebaptise the republic's volcanic island-colony. Dramatic licence, of course, allows Maréchal to go further in transforming the national brotherhood that had deposed the last of the Bourbons in France into a universal (i.e. pan-European) republican league that rids the continent of all of its crowned heads, including the Spanish Bourbon king bewailing his kinship with the executed Louis XVI.[39]

This remains simply speculation. What is more certain is that Maréchal's play popularised the association of Montagnard politics and volcanic power. As the Terror took hold in the last three months of 1793 a zealous dechristianization campaign converted former churches into Temples of Reason. In Strasbourg on 20 November 1793 the revolutionaries set up a mock volcano in a former church, complete with loose rocks and shattered crater, as part of their celebration of the Festival of Reason.[40] In Paris Charles-Louis Lesur's patriotic comedy *La Veuve du républicain, ou le Calomniateur* premiered to great acclaim in late November 1793. It closes on Beauval's rousing line: '[France] is on the march and, like a volcano, will devour all [its massed enemies].'[41] A month later a further stage-volcano erupted in Toulouse, offering local audiences a thinly veiled rewriting of *Le Jugement dernier des rois*. This was citizen Desbarreaux's *Les Potentats foudroyés par la Montagne et la Raison* which gave a Rousseauist twist to Maréchal's play by making the desert island home to a marooned 'Julie' and her daughter 'Emile'. More significantly, its regicide volcano is presented as an agent not only of revolutionary nature but also of atheist reason.[42] A more abstract use of volcanic scenery and metaphor opened in Paris on 17 January 1794 at the Théâtre de la

38. For a brief historical overview of the island and further references, see Yvan Combeau, 'De Bourbon à La Réunion, l'histoire d'une île (du XVIIe au XXe siècle)', *Hermès* 32-33 (2002), p.91-99.
39. Maréchal, *Le Jugement dernier des rois*, p.1322-23.
40. Perović, *The Calendar in Revolutionary France*, p.166.
41. Charles-Louis Lesur, *La Veuve du républicain, ou le Calomniateur* (Paris, Crapelet, 1793), p.50.
42. Citoyen Desbarreaux, *Les Potentats foudroyés par la Montagne et la Raison, ou la Déportation des rois de l'Europe* (Toulouse, P. Francés, 1793). The play involves a

Cité-Variétés. Aristide Valcour's opera buffa *Les Petits Montagnards* mixed the volcanic landscapes of Auvergne ('le Puy de Montagnard') with yet more Rousseauist child-heroes and clear Jacobin political sympathies. Its closing song evokes the political 'Mountain' as 'the most terrible of volcanoes' that has annihilated the enemies of the Revolution, presenting it interestingly as Danton standing unmoved among an agitated throng, 'the unshakable Mountain' that makes all its adversaries cower and tremble.[43] What we see here is the theatrical use of the volcano changing to chart the politics of the early Terror in which a violently anti-clerical reason is the presiding virtue and the influence of Rousseau is most strongly felt in the identification of the Revolution with nature, albeit a godless one.

Yet from early February 1794 a second phase of the Terror, overseen by Robespierre, denounced atheist radicals as aristocrats and foreign spies, promoting instead a more faithfully Rousseauist cult of the Supreme Being. This shift in revolutionary ideology is reflected in François Cizos-Duplessis's five-act 'dramatic Allegory' entitled *Les Peuples et les rois, ou le Tribunal de la raison* which was first performed in Paris on 12 April 1794. Its final act opens in a desert decor in the middle of which rises a 'rock whose summit intermittently spouts flames'. The scene is thus set for a villainous conspiracy hatched by a 'false patriot', an Englishman and a German. As the Englishman hands out money to the French traitor in order to bribe and corrupt his countrymen, he finds the air increasingly unbreathable, fearing that 'this volcano might explode' at any moment. The Frenchman also remarks that 'this rock has never vomited such flames before', but consoles himself with the thought of the strife and anarchy that he will sow in France with the help of his foreign accomplices. It is tellingly when the traitor gloats about the counter-revolutionary 'masterstroke' of divisive, enervating 'federalism' that the revolutionary volcano finally erupts, wiping out the conspirators and blowing apart both rock and desert scenery.[44] In their place a new mountain rises in an Arcadian landscape, peopled by revolutionary virtues, bearing the inscription 'Temple of Nature'. And it is from this mountain that

double perversion of Rousseau's *Emile* (1762), since it changes the eponymous student's gender and replaces his tolerant theism with radical atheism.

43. Aristide Valcour, *Les Petits Montagnards, opéra-bouffon* (Paris, Cailleau, 1794), p.64. The music was composed by Citoyen Foignet.

44. François Cizos-Duplessis, *Les Peuples et les rois, ou le Tribunal de la raison* (Paris, Barba, 1794), p.51-54.

revolutionary Reason speaks to her people; but this is now a Reason fully reconciled to deist religion as well as to Rousseauist nature. The destructive volcano that blasted the sneaking counter-revolutionary plotters is replaced by a mountain topped by Reason holding a burning torch. Reason's torch is subsequently handed to a virtuous sans-culotte so that he might light an 'altar of Humanity' and burn on it the contemptible relics of former kingship and priesthood to the greater glory of the Supreme Being, guarantor of all popular virtues.[45]

As a literary and theatrical figure, the volcano thus articulated and negotiated the shifts in extremist revolutionary politics in 1793 and 1794. Yet, due to its close association with the Terrorist regime, it necessarily fell with Robespierre and his allies in July 1794. As the nineteenth century dawned, theatrical revivals of volcanic decors thus tended to be limited to spectacular, explosive *coups de théâtre*, as in Thomas Dibdin's pantomime *The Volcano, or the Rival harlequins* (1799) staged at the Theatre Royal in Covent Garden in which the volcanic daemon of the mountain, Cratero, is thwarted in his villainous plans by Floridel, the nimble spirit of the air.[46] In France, where the memory of the Terror haunted subsequent generations, the device of a violently erupting volcano on stage was only to be successfully revived as late as 1827, and only then in the romanticised context of Neapolitan banditry in René-Charles Guilbert de Pixerécourt's melodrama *La Tête de mort, ou les Ruines de Pompeïa*.[47] We therefore need to return to the volcano and the revolutionary Terror to understand more fully the deep and resonant political imbrication of the one with the other in the Revolution.

The volcano and the Terror

On 19 July 1793 the champion of women's rights and Girondin sympathiser Olympe de Gouges stuck a yellowish poster on the walls of Paris. It was an anonymous tract entitled *Les Trois Urnes, ou le Salut*

45. Cizos-Duplessis, *Les Peuples et les rois*, p.54-56, 61.
46. Thomas Dibdin, *Songs, recitatives, &c. in the volcano, or the Rival harlequins* (London, J. Barker, 1799).
47. René-Charles Guilbert de Pixerécourt, *La Tête de mort, ou les Ruines de Pompeïa* (Paris, 1828). Pixerécourt had, however, experimented with melodramatic volcanic backdrops as early as 1818 in his *Belvéder, ou la Vallée de l'Etna*. The scenery was masterfully created by Louis Daguerre and, especially in the third and final act, Etna intermittently rumbles and spits flames to reflect the plot's heightened tensions. See *Belvéder, ou la Vallée de l'Etna* (Paris, 1818), p.44, 55.

de la patrie, par un voyageur aérien and it argued for a nationwide truce to allow for a month-long plebiscite of all citizens so that they might choose by ballot once and for all between three forms of regime: a unitary Jacobin republic, a federative republic or a monarchy.[48] The poster was illegal, as the Convention had decreed in March 1793 that only a centralised 'one and indivisible' form of Jacobin republicanism was admitted in revolutionary France. All other forms of government were outlawed and any promotion of them was punishable by death. It was, in short, a seditious bill designed principally to challenge the newly adopted Montagnard Constitution and to advocate Girondin federalism in its stead. It is of interest to us because de Gouges's poster counters the volcanic rhetoric of the revolutionary Mountain with an alternative image of sublime heights. Citing three lines from Alexis Piron's comedy *La Métromanie* (1738), de Gouges appeals to her fellow citizens to have enough strength to reject the self-interested factions tearing France apart, since

> What can a rolling wave do against a rock?
> Was Hercules bound by all the Pygmies' work?
> Olympus gazes peacefully on Mount Etna's smoke.[49]

The provocation of the last line is clear: Olympe/Olympus defies all the noise and fury of the revolutionary Mountain/Etna; the lofty abode of the Greek gods affords a measured alternative to all the discordant smoke and thunder of the Jacobins. Yet this Olympe was to fare no better than Sade's unfortunate Olympe de Borghèse in taking on the power of the volcano. Betrayed to the pro-Montagnard authorities by her own printer, de Gouges was arrested on 20 July 1793 and spent the next three months incarcerated in a succession of overcrowded revolutionary jails. She was eventually brought before the Revolutionary Tribunal on 2 November and followed her Girondin heroes to the guillotine the following day, meeting her death with courage and dignity.[50]

De Gouges did not fall victim to revolutionary Terror just because she backed the wrong side in the deadly power struggle pitting the

48. Olympe de Gouges, *Les Trois Urnes, ou le Salut de la patrie, par un voyageur aérien*, in *Ecrits politiques*, 2 vols (Paris, 1993), vol.2, p.243-48.
49. De Gouges, *Les Trois Urnes*, p.244. See Alexis Piron, *La Métromanie, ou le Poète*, in *Théâtre du XVIIIe siècle*, ed. J. Truchet, vol.1, p.1015-1131 (1080).
50. See Olivier Blanc's 'Introduction' in Olympe de Gouges, *Ecrits politiques*, vol.2, p.7-42.

Gironde against the Montagne. She also underestimated the pervasive power of Jacobin discourse in the summer of 1793 when, as Mary Ashburn Miller describes it, 'the simultaneous politicization of the natural world, and the naturalization of political rhetoric' took hold.[51] To take the example of the citation above, de Gouges misunderstood that the volcano was no longer a figure of dangerous volatility and wanton destructiveness, to be negatively contrasted with the sublime order reigning on Olympus. Rather, from July 1793 onwards the volcano was emblematic of spontaneous outbursts of patriotic passion and the symbol of an implacable purifying power. In other words, it was set to become an important valorising marker of revolutionary Terror. If we take the Terror to constitute the ruthlessly radical phase of revolutionary politics unfolding from September 1793 to July 1794, then it began with a volcanic call to arms. Among the important appeals issued on 5 September 1793 to make 'terror' the governmental 'order of the day', a particularly impassioned plea came from the president of the revolutionary Commune of Paris (effectively the municipal council). In the context of worsening civil strife, serious military defeats and soaring food prices, Pierre Gaspard Chaumette denounced false republicans, heartless food-hoarders and factious plotters as 'reptiles who corrupt everything they touch', asking rhetorically: where is 'the powerful being whose terrible cry will awaken slumbering, or rather paralysed, justice'? He then addressed the far-left Jacobin deputies directly:

> And you, Mountain, [...] be the 'Sinai' of the French! Amid lightning bolts, hurl the eternal decrees of justice and the will of the people! [...] For long enough the concentrated fire of the love of public good has coursed hotly through your flanks, let it now erupt violently! Sacred Mountain! Become a volcano whose burning lava destroys forever the hopes of the vicious and turns to ashes hearts still harbouring the idea of royalty![52]

The volcano represents not just revolutionary benevolence and righteousness (public good and the rule of law), but also the passions that authenticate this politics. As Chaumette went on to declare, amid rapturous applause, 'the day of justice and anger has come!'[53] The volcano is not just the figure of swift purgative revolutionary law in action, it also embodies the boiling, eruptive emotions which justify it.

51. Ashburn Miller, *A Natural history of revolution*, p.140.
52. Madival, Laurent *et al.*, *Archives parlementaires*, vol.73, p.410.
53. Madival, Laurent *et al.*, *Archives parlementaires*, vol.73, p.410.

This is particularly apposite in the politics of the Terror. As the name suggests, the Terror is as much a politics of mass psychological intimidation as it is one of exemplary physical violence. Among the conceptions of terror familiar to late-eighteenth-century European societies, there were three primary plausible models: the dynamic political principle of despotism, fear (*crainte*), as defined by Montesquieu; the intense fright (*frayeur*) induced by tragic theatre, as a counterpart to pity, in the Aristotelian theory of dramatic catharsis; and the paralysing dread (*terreur*) seizing on those faced with overwhelming natural forces or boundless expanses that threatened to annihilate them – otherwise perceived as 'sublime' from a safe remove.[54] The revolutionary Terror played effectively on all of these keys. Yet in its increasing self-identification with the irresistible, terrifying power of nature's 'revolutions' – hurricane, earthquake, tidal wave, volcanic eruption, that is, the most elemental forces of air, earth, water and fire – the politics of the Terror was more determined by this last conception of terror than by other possible models. One might even argue, as Ashburn Miller does, that the Terror was *overdetermined* by the 'naturalising' language it used to illustrate its unbreakable resolve and merciless implementation of policy.[55] Moreover, because the legitimacy of extreme revolutionary action, hitherto grounded in the *peuple*, was thus subtly displaced onto extreme geological or climatic phenomena, it heightened a sense of fatalism in both its victims and its executive. The earthquake and the volcanic eruption are at once unpredictable and inevitable, so the politics modelled on them must be similarly unpredictable and inevitable; therefore, why try to understand or to resist? This is providence in reverse: the revolutionary people and its government act now in the service of nature, not the other way round; their discourses and their actions are the inscrutable means ultimately serving sovereign nature's ends. The greatest revolutionary good can then only come from a self-abnegating, affective merger with the natural powers operative in revolutionary society. The Terror, as Burke had foresaw, is an Empedoclean not a Promethean revolution.

In the initial dechristianising phase of the Terror anti-clerical revolutionaries were aware of the atavistic character of popular fear. That is, they made the assumption that the most intensive and effective

54. For Montesquieu's notion of despotism's rule of fear, see *De l'esprit des lois* (1748); for Aristotle's notion of cathartic fright, see *Poetics* (*c.*335 BCE); for natural forms of terror, see chapter 3 in the present volume.
55. Ashburn Miller, *A Natural history of revolution*, p.14-16.

form of fear drew on our ancestral memory of terrifying encounters with inconceivably powerful natural forces such as storm-tides and volcanic eruptions. Ironically, they recognised that this was also the source of primitive religious beliefs, when the first self-appointed priesthoods enslaved frightened populations with their mystifying interpretations of devastating natural phenomena.[56] Yet the radical revolutionaries of late 1793 did not want to ward off storms and volcanoes, but to hail their incommensurable material power and to identify with it. As Maréchal's play makes clear, the volcano, just like the dechristianising Terror, abhors organised religion and engulfs defenders of Catholic and Protestant faiths alike in boiling lava. The volcano is an active agent of atheist Revolution; it does not want to dissipate terror but to incarnate it. As a result, the volcano alone is worthy of revolutionary veneration – 'sacred Mountain!' as Chaumette declared.

Nonetheless, even when the deist faction of Robespierrists supplanted and guillotined the leading dechristianisers in February-March 1794, the volcano remained a potent icon of Terrorist ideology. But it was now to be exploited less materially and more symbolically. This emerged clearly in the spate of revolutionary festivals of natural religion held through the spring of 1794, culminating in Robespierre's famous Festival of the Supreme Being. Thus on 9 May, just two days after the 'Incorruptible' had declared the cult of the Supreme Being to be the new state religion, the district of Bellevue-les-Bains in central-eastern France organised a parade of busts of revolutionary heroes alongside a portable mountain bearing the inscription 'From my summit emerges the lightning that pulverises the enemies of the people and the republic' (the nearby town of Motte-Saint-Jean had similarly been secularised as Montagne-Fleurie in 1793).[57] Yet the ultimate consecration of mountainous revolutionary iconography came with the monumental Festival of the Supreme Being in Paris on 8 June. As Marie-Hélène Huet has convincingly argued, this was one of the fundamental initiatives taken by Robespierre in order to inaugurate a sublime revolutionary Republic of Virtue.[58] This meant eliminating as

56. This was a thesis dear to the atheist materialists like Paul-Henri Thiry d'Holbach and Nicolas-Antoine Boulanger. See Perović, *The Calendar in Revolutionary France*, p.80.

57. See Ashburn Miller, *A Natural history of revolution*, p.133.

58. Marie-Hélène Huet, 'Le sacre du printemps: essai sur le sublime et la Terreur', *Modern language notes* 103:4 (1988), p.782-99.

far as possible all profane representations of the Revolution. To inspire his fellow citizens to self-sacrificing virtue, Robespierre thus sought to impose on the republic a deist religion of the unrepresentable sublime, invoking the Rousseauist divinity of the Supreme Being to this end. Yet as he had already declared in February 1794, republican virtue was vulnerable to subversion and corruption in this time of continent-wide war; as such, virtue would prove 'powerless' to found the republic unless it was seconded by its natural ally, lethal 'terror'. Virtue was the blessed state of oneness with the Revolution; terror was an 'emanation' or expression of the violent struggle necessary to attain virtue in a world beset by counter-revolutionary treachery. Terror thus constituted the very form of the virtuous citizenry's 'prompt, severe, inflexible justice' against all perceived forces of reaction.[59] Put more prosaically, virtue was the revolutionary carrot; terror, the revolutionary stick. So at the same time as the Champ de Mars (renamed the Champ de la réunion) was being landscaped into a tiered peak for the festival, the Committee of Public Safety was formulating the so-called Law of 22 Prairial (voted 10 June 1794) which drastically streamlined the judicial process, stripping the Revolutionary Tribunal of its political theatre and denying the accused the right to any legal representation at all. The only possible verdicts were to be acquittal or death – but mainly the latter. The guillotine was simultaneously removed as a gory public spectacle in central Paris and resurrected in the eastern suburbs of the capital where it set about its work more quietly and more efficiently. Robespierre's idea in both instances – for the Festival of the Supreme Being and the Law of 22 Prairial – seems to have been to elide as far as possible all notions that the sublime Revolution could be 'represented' by material means either in its deist faith or in its terrible justice.

Yet the manifest paradox was that the Festival of the Supreme Being relied precisely on an accumulation of revolutionary symbolism in order to realise the cult designed to transcend all profane symbols of Revolution.[60] Chief among these symbols was the mountain, and by extension the volcano. In fact, the festivities began in the Tuileries gardens with a large bonfire. Robespierre, acting as president of the National Assembly, and dressed in a sky-blue suit, lit an enormous pyre in the form of a gigantic allegory of Atheism. If it appeared to burn

59. Maximilien Robespierre, *Œuvres de Robespierre*, 10 vols (Paris, 1912-1967), vol.10, p.350-67 (357).

60. See Mona Ozouf, *Festivals and the French Revolution*, translated by Alan Sheridan (Cambridge, MA, 1988), p.111-18.

more swiftly and intensely than usual this was because its creator, the pyrotechnical showman Petrone Ruggieri, had treated the monstrous effigy with inflammable chemicals to accelerate combustion.[61] Thus Godlessness roared into flames, consuming at the same time the allegorical figures of attendant evils Ambition, Egotism, Discord and Hypocrisy. In their place rose a slightly sooty, but none the less dignified, statue of Wisdom. Then the entire Convention processed in rather unruly fashion to the Champ de la réunion as the massed choir struck up the 'Hymne à l'Etre Suprême'. There Robespierre ascended the rocky, sculpted mountain at its centre surrounded by floral decorations, and before an estimated crowd of 100,000 people celebrated the union of the Revolution and nature, conjointly blessed by the Supreme Being. In terms of its particular volcanic attributes, the bulking mountain constructed under the direction of the painter Jacques-Louis David bore large incense-burners on all flanks that smoked like fumeroles and at its craggy peak a liberty tree spread its branches like a Plinian plume of ash.[62] At least one artist, inspired by this rocky structure, drew a direct parallel between the mountain of the Supreme Being and an active volcano. An anonymous coloured engraving of the Festival, seen from 'the eastern side of the mountain', carries the verse-legend:

> At the imposing sight of the proud Mountain,
> Tremble audacious tyrants…!
> It crushes all the factions
> As the volcano scours the land in flame.[63]

In this way the associations of the political Mountain of the Convention and the erupting volcano of Revolution remained much the same under the benign gaze of the Supreme Being as under the godless skies of the early Terror in late 1793.

61. See Luce-Marie Albigès, 'La fête de l'Etre Suprême, 20 prairial An II (8 juin 1794)', http://www.histoire-image.org/etudes/fete-etre-supreme-20-prairial-ii-8-juin-1794 (last accessed 14 December 2018).
62. For full details of the constructions and unfolding of events at the *fête de l'Etre Suprême*, see Jonathan Smyth, *Robespierre and the Festival of the Supreme Being: The search for a republican morality* (Manchester, 2016), p.49-77. Smyth gives an estimated height for Robespierre's platform of 40 feet above crowd level (p.59).
63. 'Vüe du côté oriental de la montagne élevée au Champ de la réunion pour la fête célébrée en l'honneur de l'Etre Suprême' (1794), Paris, Bibliothèque nationale de France, De Vinck, 6306. Reproduced in Ashburn Miller, *A Natural history of revolution*, p.137.

Yet if the Festival of the Supreme Being is sometimes considered the pinnacle of Robespierre's personal power in the Revolution, the overt references here to 'audacious tyrants' and the 'proud Mountain' were implicit reminders that he was also increasingly caricatured as a murderous megalomaniac bent on dictatorship, posing a real threat to parliamentary rule. It also recalls the fact that the Terror itself was far from being the inevitable outcome of irreversible, escalating revolutionary brutality, even if evocations of volcanoes, earthquakes and hurricanes implied a diminished human agency in popular violence and made it appear more 'natural', predetermined and fateful than it actually was. Despite the revisionist and post-revisionist historiography of François Furet and his followers, the Terror was not intrinsic to the Revolution or an inescapable consequence of the political violence beginning with the first popular lynchings of 1789.[64] It is, moreover, one of the peculiarities of the highly malleable figure of the volcano in the Revolution that it can embody these antithetical ideological interpretations. That is, the volcano can stand just as aptly as a marker of the dizzying contingency of revolutionary politics as it does of the Revolution's irresistible might or ineluctable course. A good example of this would be the astute reflections on the Terror provided by the Englishwoman Helen Maria Williams, a friend of the Girondins, arrested and incarcerated in the Luxembourg prison in late 1793. Recalling the acute uncertainty of day-to-day existence under the Terror on the eve of her arrest, she writes in her *Letters containing a sketch of the politics of France* (1795):

> The political clouds in the meantime gathered thick around the hemisphere: we heard rumours of severity and terror, which seemed like those hollow noises that roll in the dark gulph of the volcano, and portend its dangerous eruptions: but no one could calculate how far the threatened mischief would extend, and how wide a waste of ruin would desolate the land.[65]

The volcano of revolutionary politics rumbles but does not yet erupt, at the very time that Maréchal, Chaumette and others are imagining

64. See, for instance, François Furet, *Penser la Révolution française* (Paris, 1978), and Simon Schama, *Citizens: a chronicle of the French Revolution* (London, 1989).
65. Helen Maria Williams, *Letters containing a sketch of the politics of France, from the thirty-first of May 1793, till the twenty-eight of July 1794, and of the scenes which have passed in the prisons of Paris*, 2 vols (London, G. G. and J. Robinson, 1795), vol.1, p.5.

torrents of lava consuming the impotent enemies of the republic. What Williams' volcano registers instead is the sheer precariousness of living on the slopes of the smoking, grumbling mountain; she thrusts us into the daily unpredictability of revolutionary politics as experienced by an attentive witness to the tumultuous, confused decisions and events of the period.[66]

Terrorist politicians also fell back on this idea of the volcano to convey the treacherous latency of counter-Revolution. This is exemplified in Antoine-Louis de Saint-Just's prosecution of the so-called Dantonists delivered to the Convention on 31 March 1794, just a week after the Jacobin deputy had successfully dispatched the far-left Hébertists to the guillotine. Its tone is set with the famous declaration: 'There is something terrible in the sacred love for the fatherland; it is so exclusive that it sacrifices everything without pity, without fear, without respect for civility, to the public interest.'[67] Saint-Just weaves a seamless narrative of hypocrisy, perfidy and treachery linking together all the defeated factions of the Revolution thus far – Orléanists, Girondins, Hébertists, Dantonists – as agents of foreign powers and internal plots working tirelessly to restore monarchy to France. In this context of seething suspicion and fear, the volcano no longer represents the explosive triumph of revolutionary republicanism but the covert rumblings of reactionary intrigues fomented from abroad. 'The people's revolution', writes Saint-Just, 'was the surface of a volcano of foreign conspiracies.'[68] His conception of the volcano of Revolution, then, coincides more with Williams' fearfulness than with his fellow Montagnards' ebullient self-image.

As the Terror reached its bloody denouement in July 1794 this negative figure of the volcano as a metaphor for imminent cataclysm became more pervasive, especially among the increasingly polarised members of the two executive Committees, that of Public Safety and that of General Security. It lent itself to imagining the eventual violent eruption and deadly fallout of their bitter internal strife. It is a catastrophic image which particularly plagued the hard-line

66. On the daily existential dilemma of surviving the Revolution, and especially the Terror, see Marisa Linton, *Choosing Terror: virtue, friendship, and authenticity in the French Revolution* (Oxford, 2013).

67. Antoine-Louis de Saint-Just, 'Rapport au nom du Comité de salut public et du Comité de sûreté générale sur la conjuration ourdie depuis plusieurs années', in *Œuvres complètes*, p.706-37 (706).

68. Saint-Just, 'Rapport au nom du Comité de salut public', p.710.

Jacobin member of the Committee of Public Safety Jacques-Nicolas Billaud-Varenne. As he perceived ideological and personal tensions reaching breaking point in July 1794 Billaud-Varenne grew moody, distrustful, withdrawn and prone to feigning sleep in meetings, and was frequently overheard muttering: 'We are walking on a volcano.'[69] Yet the definitive, fatal rift was precipitated by Robespierre himself who returned to the Convention on 26 July 1794 after six weeks' absence and gave a pained and personal, disjointed and menacing speech lasting for about two hours. He billed it as his political testament, bequeathing 'the terrible truth and death' to his enemies while also threatening Socratic suicide by hemlock.[70] Devastatingly, he called for a renewed purge of both revolutionary committees and the Jacobin Club. Yet he did not give a list of individuals to be removed, letting an ominous general accusation hang in the air of the packed Convention. In his denunciation of fickle, self-serving revolutionaries out to dupe the people, Robespierre nonetheless darkly cited Billaud-Varenne's volcanic refrain: 'Why do those who recently told you: "I declare to you that we are walking on volcanoes" now believe that they are treading on roses?'[71] This would have been a strong signal to many that Billaud-Varenne was in Robespierre's deadly sights and that the Terrorist volcano was anything but dormant, let alone extinct.

The figure of the volcano had already appeared earlier in the speech. This time it was evoked, as with Saint-Just and Williams, to suggest an impending explosion of revolutionary emotion and action. Specifically, Robespierre paints a picture of righteous popular indignation, stoked by fear, coming to the boil to burst out and blast the two counter-revolutionary forces of royalist intrigue and atheist perversion. He declares: 'a justified indignation, intensified by terror, was silently fermenting in all hearts; a terrible, inevitable eruption bubbled in the bowels of the volcano while little intellectuals played foolishly on its summit with great rogues.'[72] However, Robespierre prepares this volcanic scenario only to claim that, thanks to the Convention's intervention, and at his instigation, the decree of 7 May 1794 instituting the national cult of the Supreme Being averted the dreaded eruption. The equally disastrous options of submitting docilely to foreign-imposed monarchism or siding with godless extremists were

69. See Linton, *Choosing Terror*, p.258.
70. Robespierre, *Œuvres de Robespierre*, vol.10, p.542-76 (566).
71. Robespierre, *Œuvres de Robespierre*, vol.10, p.567-68.
72. Robespierre, *Œuvres de Robespierre*, vol.10, p.560-61.

thus simultaneously defeated; the virtuous people and its virtuous representatives carried the day. The volcanic image falls flat here, being one of neither imminent doom nor irresistible triumph.

Robespierre's somewhat aborted analogy concerning the volcano in his speech of 26 July 1794 could also be taken as a small sign of a much bigger psychological breakdown. However imaginary the hordes of his enemies had been before his louring intervention in the National Assembly, an important number of them were galvanised into very real action by his rambling, threatening speech. Thus brutally merciless deputies-on-mission recalled to Paris, such as Jean-Lambert Tallien, atheist extremists in the Committee of General Security, such as Marc-Guillaume Alexis Vadier, and hostile colleagues on the Committee of Public Safety, such as Billaud-Varenne and Jean-Marie Collot d'Herbois, formed a concerted plan to strike. On the following morning in the Convention Saint-Just began a calculated defence of Robespierre and the Terror. At this point, first Tallien, then Billaud-Varenne, then tens of other deputies interrupted and hundreds more drowned out Robespierre and his allies' attempts to defend themselves in a deafening roar of abuse and execration. Within two hours (the length of Robespierre's speech the day before) it was all over. The Robespierrists were under arrest, the Jacobin Club would be locked that evening, and an intense disinformation campaign tarnishing Robespierre as a would-be 'dictator' – even bizarrely as an alcoholic womaniser – stirred into action. The 'Incorruptible' and his closest political supporters were guillotined without trial on 28 July 1794.[73]

As it transpired, the interruption of Saint-Just's speech prevented him from reaching the passage where he too recalled Billaud-Varenne's volcanic doom-mongering: 'Billaud', he wrote, 'often repeats these words with feigned horror: *We are walking on a volcano.*' Like Robespierre the day before, Saint-Just brings up this verbal tic the better to accuse Billaud-Varenne of hypocrisy and personal designs on power: 'I think so too [that we are walking on a volcano]. But the volcano we are walking on is that of his [Billaud's] dissimulation and his love of domination.'[74] The seemingly compulsive manner in which Billaud-Varenne uttered this phrase about 'walking on a volcano' is thus presented as a subconscious proof of his poorly repressed urge to destroy those who get in his way, a concealed desire to rule that reveals itself unwittingly in the recurrent image of latent volcanic

73. See Linton, *Choosing Terror*, p.251-71.

74. Saint-Just, 'Discours du 9 thermidor An II', in *Œuvres complètes*, p.769-85 (774).

violence. Yet, as we know, Billaud-Varenne pre-empted this accusation in shouting down his fellow Jacobin; thus another Robespierrist volcano misfired. As Marisa Linton argues convincingly, such emotive and discursive contingencies contributed as much to the end of the Terror as did ideological differences between its protagonists.[75]

Billaud-Varenne was to have the last word in another way too. Deported to French Guiana with fellow Terrorist Collot d'Herbois in May 1795, Billaud-Varenne would live out the years of the Directoire and the rise and fall of Napoleon before dying of dysentery in 1819 in Haiti. His apocryphal *Mémoires* were published two years later, relating the former Jacobin's imaginary peregrinations in Central America. While the text is clearly a forgery, the fiery volcanic landscapes of Mexico and Guatemala inspire the fictionalised ex-revolutionary to further explosive comparisons with the events of the Revolution. To give but one example, scaling the mighty Popocatépetl, 'Billaud' claims that the Mexican mountain had erupted violently in 1789 – 'The same year a more terrible volcano opened up in France; it devoured the monarchy and its convulsions, reverberating from shore to shore, in spite of despotism, are going to change the face of the world.'[76] Yet, oddly enough, as the real Billaud-Varenne would have been only too well aware, the astonishing coincidence of a shattering volcanic eruption and extreme revolutionary politics was far from being the stuff of picaresque fiction.

June 1794: Vesuvius and the Terror

On 7 June 1794, a day before the Festival of the Supreme Being, a sudden puff of smoke burst from the earth in the upper vineyards on the western flank of Vesuvius, petrifying a peasant and two boys tending their crops. Eight days later at 10pm a spurting fountain of fire and billowing clouds of black smoke erupted from the same spot. At 2am on 16 June another lava spout issued from the eastern flank and started to snake towards Resina. A further fourteen jetting vents then opened up down the western side of the mountain, forming a roaring, red gash pouring out tons of lava in the direction of the sea.[77] As later reported by Sir William Hamilton, a gigantic discharge of

75. Linton, *Choosing Terror*, p.285-89.

76. Jacques-Nicolas Billaud-Varenne, *Mémoires de Billaud-Varennes, ex-conventionnel, écrits au Port-au-Prince en 1818*, 2 vols (Paris, 1821), vol.1, p.132.

77. Scarth, *Vesuvius: a biography*, p.212-23.

smoky ash clouds, crackling violently with zigzag lightning, rose miles into the early morning sky, taking on the ominous form of a spreading pine tree.[78] By 5am on 16 June the main torrent of churning lava, now nearly a mile wide, reached and subsumed the town of Torre del Greco. It swept through the main streets of the town, piling up against the church to a height of 40 feet and igniting timbers and gunpowder stocks in houses throughout the town. An hour later the lava hissed into the sea, boiling fish in geysers of steaming seawater and forming a cooling promontory three quarters of a mile wide, 24 feet deep and half a mile long. Of the 18,000 inhabitants of Torre del Greco, only fifteen or so elderly perished in the lava flows, although over 3000 acres of vineyards were incinerated and encased in molten rock.[79] In Naples the ashes fell several inches thick and wealthy residents protected themselves with umbrellas, while three people died in the village of Somma when a roof collapsed under the weight of ash showers.[80] The darkness was such that on 19 June candles were lit at midday in Caserta and the sun was obscured as far away as Benevento, some forty miles from Naples. To compound the devastation and misery, enormous volcanic cloudbursts in early July created a thick rain of wet ash and set off deadly mudslides, some a quarter of a mile wide, which killed over 4000 livestock and hardened into a solid natural cement on which, writes Hamilton, 'nothing less than a pick-axe could make any impression'. Finally, the eruption released noxious vapours called *mofette*, consisting largely of carbon dioxide, which hugged the ground, filled cellars, wells and natural depressions, suffocating over 1300 hare and game birds as well as taking the life of a young lad who fell into an empty well in late August 1794.[81] Many of the indigent local inhabitants refused to move to Naples or elsewhere in the region. They slowly returned to the charred remains of their houses and attempted to excavate and rebuild them, although, after a visit to Torre del Greco in 1797, Mariana Starke claimed that the lava still choked half of the town and was 'so hot in several places as to kindle a torch'.[82]

78. Hamilton, 'An account of the late eruption of Mount Vesuvius', p.81-83.
79. Hamilton, 'An account of the late eruption of Mount Vesuvius', p.81-98.
80. Starke, *Letters from Italy*, vol.2, p.136; Hamilton, 'An account of the late eruption of Mount Vesuvius', p.100.
81. Scarth, *Vesuvius: a biography*, p.221-22; Hamilton, 'An account of the late eruption of Mount Vesuvius', p.99, 103-107, 113.
82. Starke, *Letters from Italy*, vol.2, p.137.

News of the eruption – which Hamilton maintained was the third most powerful in the volcano's history after those of 79 CE and 1631 – reached western Europe about a month later. Most revolutionary news-sheets reported the eruption: the *Mercure universel* on 14 and 15 July 1794, the *Feuille de la république* on 17 July and the *Gazette nationale, ou le Moniteur universel* on 22 July.[83] Despite mistakenly reproducing an illustration of the 1631 eruption, the *Almanach du messager boiteux*, published in Bern, stuck to the facts, singling out the crater collapse on 19 June and relaying the first estimates of damage drawn up by the royal Neapolitan engineering corps of 39 dead and 15,000 homeless.[84] The widely read French-language *Gazette d'Amsterdam* reported the eruption on 25 July; it would have done so sooner but its mail had to be rerouted via Livorno because of revolutionary violence and armed conflict along the French borders.[85] A preoccupation with revolutionary events may also have coloured its language, 'concerning the terror which has gripped the city of *Naples* on 16 June due to a new eruption of Mount *Vesuvius*'. It claimed that it was impossible to imagine 'the horrors that the sight of this tragic scene presents', although it preferred to focus on the damage done to noble and religious houses, especially at Torre dell'Annunziata, a small coastal town to the south of Torre del Greco.[86]

In contrast, the British papers seemed fascinated by the extravagance of Catholic ritual in the face of the thundering eruption. The *True Briton (1793)* of 25 July 1794 reported that the parading of San Gennaro's mummified head was 'generally accompanied by upwards of 200,000 persons'.[87] The *Sun* of 15 July gave fuller vent to a mocking anti-Catholic sentiment in this context, describing the

83. Ashburn Miller, *A Natural history of revolution*, p.160; *Gazette nationale, ou le Moniteur universel* 304, 'Quartidi 4 thermidor l'an 2 (Mardi 22 Juillet 1794 *vieux style*)', reprinted in *Réimpression de l'Ancien Moniteur, seule histoire authentique et inaltérée de la Révolution française depuis la réunion des Etats-Généraux jusqu'au Consulat (mai 1789-novembre 1799)*, 32 vols (Paris, 1858-1863), vol.21, p.270.
84. See Favier, 'Les éruptions du Vésuve', p.104, 108-109.
85. Mercier-Faivre, 'Eruptions dans la presse', p.85-86.
86. *Gazette d'Amsterdam* of 25 July 1794 (its italics), cited in Mercier-Faivre, 'Eruptions dans la presse', p.97.
87. *True Briton (1793)* 491, London (Friday 25 July 1794), p.1. The *Oracle and public advertiser* had carried news of the eruption on 14 July and had also reported the religious processions of San Gennaro's relics in Naples; however, its edition a week later noted slyly that the saint had omitted to protect the royal summer house at Portici from the lava. See *Oracle and public advertiser* 18745, London (Monday 14 July 1794), p.1.

'silly bigotry [...] resumed at Naples in consequence of the recent eruption of Vesuvius'. It went on to print a doggerel by the popular verse satirist John Wolcot, writing under his pseudonym of Peter Pindar. It ran:

> When Mount *Vesuvius* pour'd his flames;
> And frighten'd all the *Naples* Dames,
> What did the timid Dames, I trow –
> Why, order'd a fat Cardinal to go
> With good St. Janarius's head;
> And shake it at the Mountain 'midst its riot,
> To try to keep the Bully quiet:
> The Parson went, and shook the jowl, and sped;
> Snug was the word; the flames at once kept house;
> The frighten'd Mount grew mute as any mouse.[88]

While these comic images of papist superstition might have amused the *Sun*'s readership, they did so partly as a consoling distraction from other news issuing from the continent, arrayed in columns surrounding the report of Vesuvius' violent outburst. This included an account of the defiant jingoistic rhetoric coming out of the revolutionary Convention on 1 July 1794, the work of the Toulouse lawyer and mouthpiece for the Terror Bertrand Barère. A wily political opportunist, and ultimately the longest-lived member of the Committee of Public Safety, Barère would have no doubt appreciated Wolcot's ability to turn a volcanic disaster into political capital. Tellingly, he had himself already resorted to cataclysmic volcanic imagery on a number of important occasions. In December 1793, anticipating the retaking of the port of Toulon from the British navy, Barère cited the atheist Terrorist Joseph Fouché who had been dispatched to stiffen the resolve of the French troops in the area. Addressing the Convention, Barère projected the apocalyptic vision of the revolutionary republic as a single volcanic eruption hurling upon the British an irresistible 'devouring lava' that would extend to engulf the very isles from which the old enemy came. His volcanic fire was once again trained on the British in May 1794 in the wake of two failed assassination attempts made against Robespierre and Collot d'Herbois. He linked these plots to the earlier 'conspiracies' of Hébert and Danton, a time when 'we were sleeping on a volcano' and the republic had risked being reduced to 'a heap of ruins and ashes' by foreign-funded treason. Thus,

88. *Sun* 560, London (Tuesday 15 July 1794), p.3.

he declared chillingly, no English or Hanoverian prisoners would henceforth be taken on the field of battle.[89]

It is not surprising, then, that Barère seized on the almighty eruption of Vesuvius in June 1794 to embellish his belligerent revolutionary rhetoric. Delivering a report to the Convention in early July designed to maximise saltpetre and gunpowder production in the republic, he cites the recent crushing victory of the republican armies at Fleurus in Belgium (26 June) and tells fellow citizens not to fritter away in celebratory firework displays the gunpowder meant to kill the enemies of the Revolution. It is nature, he proclaims, 'who has organised revolutions and ceaselessly makes them throughout the universe'. Nature 'has just united with us to ripen our harvests earlier' and 'united even more so with our armies to frighten the kinglets [roitelets] of Italy' from the heights of the Alps. He then declares: 'nature, from the summit of Vesuvius, has just commanded Naples not to let sail the fleet that was going to join the English on the Mediterranean coast'; 'the revolution of *this other mountain* hampers the projects of the Italian governments.' Having revived the Terrorist link between the volcanic mountain and the Jacobin Mountain, Barère announces that 'this coalition of liberty and nature, of the lightning bolts from Mount Cenis and the lava of Vesuvius, is easily worth more than the treaty of Pillnitz [the counter-revolutionary declaration made jointly by Austria and Prussia in August 1791].'[90]

What is perhaps more surprising is that on 11 August 1794, two weeks to the day after the execution of the Robespierrists, Barère took to the tribune of the Convention and grandiloquently indulged a further comparison between an erupting volcano and the Jacobin Mountain. Republican troops had just stormed the royalist northern Spanish city of Saint-Sebastien, and in announcing its humiliating capitulation, he attempted to blast at once France's foreign foes and his former colleagues on the Committee of Public Safety in his sweeping volcanic rhetoric:

89. Bertrand Barère, *Rapport du citoyen Barère, sur la reprise de Toulon par l'armée de la République* (Paris, Imprimerie nationale, 1793), p.7; *Rapport sur les crimes de l'Angleterre envers le peuple français, et sur ses attentats contre la liberté des nations* (Paris, Imprimerie nationale, 1794), p.8. Also cited in Ashburn Miller, *A Natural history of revolution*, p.149, 14-15.

90. Bertrand Barère, *Rapport fait au nom du Comité de salut public, sur l'état de la fabrication révolutionnaire du salpêtre et de la poudre* (Paris, Imprimerie nationale, 1794), p.18, original emphasis. Also cited in Ashburn Miller, *A Natural history of revolution*, p.161.

> The republic is a concerted volcano which overthrows both haughty cities and guilty individuals, which destroys in turn supporters of foreign powers and accomplices of tyranny, which sweeps away in the same instant petty intriguers and the grandly ambitious, which is able to vanquish kings' armies and dictators' henchmen. Its eruption is equally violent internally and externally; its lava is just as burning at the centre as at the extremes; and this volcano also has its Mountain to rain fire on insolent nobility, on venal plots, on hypocritical ambition, and on all private passions; fire must purge everything and respect only the love of the fatherland.[91]

As Barère was to discover, however, the time of radical Jacobin volcanoes was over. He had long argued that intensive nationalist indoctrination and an all-consuming 'love of the fatherland' should be the only state religion. His attempt in early August 1794 to rearticulate this belief in similarly all-consuming volcanic language relied on a trope and an association with extremist Jacobinism that were ill-judged. Far from deflecting attention from his former collaboration with Robespierre and the Terrorist Committee of Public Safety, the figure of the fire-spewing volcano renewed it. Sentenced to deportation in spring 1795 with Billaud-Varenne and Collot d'Herbois, Barère escaped transportation to French Guiana but was to spend the rest of his long life under the constant threat of execution, imprisonment or exile.[92]

For the enemies of revolutionary France the conflation of volcanic and political violence in the summer of 1794 seemed more than simply rhetorical. In the aftermath of Vesuvius' destructive eruption in June, eyewitness accounts of its devastation were often materially interwoven with verbatim reports from the bellicose French National Assembly. A good example of this troubling juxtaposition of havoc-wreaking volcanism and aggressive republicanism is found in London's *Star* news-sheet of Friday 18 July 1794. On its first full page of news, two columns ran adjoining stories of 'Italy. Eruption of Mount Vesuvius' and 'France. Society of the Jacobins'. The fiery anti-British speeches of Georges Couthon and Robespierre vie for the reader's attention with a terrifying eyewitness account of 'the grandest, the most awful

91. Bertrand Barère, cited in the *Gazette nationale, ou le Moniteur universel* 326, 'Sextidi 26 thermidor l'An 2 (Mercredi 13 Août 1794 *vieux style*)', reprinted in *Réimpression de l'Ancien Moniteur*, vol.21, p.460.
92. Bertrand Barère, *Mémoires de B. Barère, membre de la Constituante, de la Convention, du Comité de salut public, et de la Chambre des représentants*, 4 vols (Paris, 1842-1844).

sight, I ever beheld'. On the one hand, Couthon urges 'we must adopt a resolution to exterminate all those who are adverse to the Republic'; on the other, 'the lava broke out on many places on the sides and, it is said, from thirty-six different openings: these, soon uniting in one great river of fire, flowed with great rapidity towards Torre del Græco.'[93] Readers would be forgiven for imagining converging streams of raging sans-culottes and red-hot lava.

In this confusion of volcanic and revolutionary news, as we have seen with Wolcot, satire was often an immediate riposte and a collective defence mechanism. In Britain in the 1790s, none was more aware of this than the caricaturist and printmaker James Gillray. Initially favourable to the Revolution in 1789, Gillray shifted to open opposition to the new regime in France with Louis XVI's unsuccessful flight to Varennes in June 1791. By 1794 he had produced a range of scathing caricatures of French republicanism, emphasising its blood-lust, impiety and emaciated fanaticism, playing on established anti-Gallic and anti-Catholic tropes. His targets were just as much, if not more, the home-grown radicals who supported the increasingly bloody Revolution as they were the French leaders and sans-culottes.[94] Thus shortly after news of Vesuvius' terrible eruption reached Britain, Gillray published his allusive and brilliant *Eruption of the mountain, or the Horrors of the "Bocca del Inferno"* (Figure 6). This hand-coloured etching appeared in his printer Hannah Humphrey's shop window on 25 July 1794. At its centre stands Vesuvius in full fiery spate, running with lava and billowing out grey ash clouds. Yet its eruption is not a random explosion, but a series of concerted volcanic attacks on the monarchist parties of the First Coalition against the French republic exposed on the plain below. Catholic Rome is struck with a bright zigzagging thunderbolt while heavy ejecta bombard Vienna and Berlin. The low-lying battlegrounds of Flanders and Holland, represented by diminutive windmills, are surrounded by smoky flows of lava and begin to blaze. More worryingly still, another torrent

93. *Star* 1845, London (Friday 18 July 1794), p.1.

94. John Cooper, 'James Gillray and the French Revolution', *RSA journal* 137 (September 1989), p.646-51; for more on Gillray's use of national stereotypes, see Sophie Loussouarn, 'Gillray and the French Revolution', *National identities* (June 2015), p.1-17; for a French perspective, see Pascal Dupuy and Caroline Bourgeois, 'Les représentations de la République française à travers les gravures satiriques (1792-1799) et la presse quotidienne anglaise (1792-1793)', *Annales historiques de la Révolution française* 289 (1992), p.367-81 (on *The Eruption of the mountain*, p.378-79).

Figure 6: James Gillray, *The Eruption of the mountain, or the Horrors of the "Bocca del Inferno", with the head of the protector Saint Janarius carried in procession by the cardinal archevêque of the Lazaroni* (1794).

of orange lava cascades down the mountainside and surges towards London, outlined by St Paul's Cathedral and St James' Palace. At the same time an odd procession emerges from the right: it is composed of a motley of pro-French Whig politicians in a double parody of a Neapolitan rite and a Parisian mob. Richard Brinsley Sheridan holds aloft the head of Charles James Fox as a travesty of San Gennaro's relics paraded through Naples to ward off the eruptions of Vesuvius. The cartoon's subtitle dubs Sheridan 'cardinal archevêque of the Lazaroni' as a reference to his role as high priest of British radicalism, while his minority Whig colleagues are decried as populists appealing to the lowest orders of society, expressly confusing Neapolitan *lazzaroni* with Parisian sans-culottes. The unpatriotic collusion of this ragged bunch of Whigs with revolutionary France is indicated by the shared symbols of fire: the earl of Lauderdale carries a conspirator's lantern which projects a further stream of republican fire upon London while

Lord Stanhope brings up the rear holding two flaming fascicles that replicate in miniature the volcano's jetting blaze. Far from seeking to repel the volcano's outpourings, this grotesque Foxite procession, all eyes heavenward, appears to revel in the impending conflagration of London and its anti-republican allies. In some hand-coloured impressions of this caricature the perfidious political conspiracy between the revolutionary French volcano and radical British Whigs is underscored by both the spouting lava and Sheridan's robes being striped in French tricolours.

In some ways Gillray's *Eruption of the mountain* operates a complete pictorial inversion of Joseph Wright's *Eruption of Mount Vesuvius with the procession of St. Janarius's head* (1778). Whereas in Wright's painting the religious parade is a dwarfed and indistinct feature in the dim foreground, Gillray makes his troupe of grotesque Whigs the principal source of meaning in the picture. Both place the volcano in the centre of their artwork. But Vesuvius for Wright is an object of louring, orange sublimity outstripping the imagination of those beholding its eruption. For Gillray Vesuvius is first and foremost a political cipher for the Jacobin Mountain used to incriminate and ridicule Whig sympathisers with the Revolution. Scales are reversed: from volcanic to human, from natural to political, from aesthetic to ideological. Thus the summer of 1794 stands as a high point in the revolutionary politicisation of the figure of the volcano, rendered more vivid and urgent by the spectacular eruption of Vesuvius in June.

However, with the execution of the Robespierrists and the end of the Terror, the imagery of volcanism began to recede in political and social discourse. As Mary Ashburn Miller has suggested, from late 1794 onwards in France the volcano reverted to being a watchword for the twin scourges of political fanaticism and social anarchy; it was a prompt for wariness about renewed revolutionary radicalisation.[95] Elsewhere in Europe, especially across the Channel, as contemporary observers tried to take stock of the Revolution thus far, the metaphor of the volcano continued to loom large in the collective imagination. In 1795 Charles Hereford published a three-volume *History of France, from the first establishment of that monarchy, brought down to, and including a complete narrative of the late Revolution.* In the penultimate chapter of his final tome, he evoked the 'dreadful eruption from Mount Vesuvius' of June 1794, 'which far exceeded any that had been known in the memory of man', describing pathetically how 'the hopes of the peasant and the

95. Ashburn Miller, *A Natural history of revolution*, p.162.

wealth of the husbandman were in a moment blasted and destroyed' by a 'stream of liquid fire'.[96] As in Gillray's print, the natural volcano appeared at this moment to be in league with the political volcano of revolution; or as Hereford puts it rather apocalyptically, 'the elements seemed to conspire with man in spreading misery and devastation throughout the human species.'[97] Thus well into the late 1790s in Britain the French Revolution gave off sulphurous fumes and intermittently flared in volcanic fashion, as is evidenced in the correspondence of the likes of Hester Piozzi and Adam Ferguson.[98] Yet, in part due to the rise of Napoleon Bonaparte from 1799 onwards, the nineteenth century deployed an increasingly depoliticised image of the volcano, one that clung more to the smouldering characters that populated Romantic and Gothic fiction than to the explosive collective forces animating European politics.

96. Charles John Ann Hereford, *The History of France, from the first establishment of that monarchy, brought down to, and including a complete narrative of the late Revolution*, 3 vols (London, C. & C. Kearsley, 1795), vol.3, p.195-96.
97. Hereford, *The History of France*, vol.3, p.196.
98. See Hester Lynch Piozzi to Reverend Leonard Chappelow, Friday 30 September 1796, Tremeirchion, Wales. *Electronic Enlightenment scholarly edition of correspondence*, http://dx.doi.org.sheffield.idm.oclc.org/10.13051/ee:doc/piozheUD0020377a1c (last accessed 14 December 2018); and Adam Ferguson to Sir John Macpherson, Thursday 21 February 1799, Hallyards, near Peebles, Scotland. *Electronic Enlightenment scholarly edition of correspondence*, http://dx.doi.org.sheffield.idm.oclc.org/10.13051/ee:doc/fergadPC0020448a1c (last accessed 14 December 2018).

6. Volcanic Iceland: conquering Hekla and surviving Laki

Iceland is Europe's most volcano-rich country, composed almost totally of volcanic basalt and boasting well over one hundred active and extinct craters. It owes its volcanism to its geography, sitting astride the Mid-Atlantic Ridge, the boundary at which the huge Eurasian and North American tectonic plates diverge. The ridge cleaves the island of Iceland in half and is one of only two places on dry land where major tectonic plates can be observed drifting apart (the other is the Great Rift Valley in East Africa). The rate of drift is as much as 2.5 centimetres per year.[1] In 930 CE the first settlers in Iceland instituted their ancient parliament, the Alþingi (Althing), directly on the ridge as it passes through Þingvellir in the south-east of the island. However, the Icelandic Commonwealth founded there was to dissolve in internal strife and be subjugated to the Norwegian crown in 1262. Exactly 400 years later it was integrated into the absolutist kingdom of Denmark–Norway. Hence by the eighteenth century, as an institution of an outlying dependency of the Danish crown, the Alþingi convened only symbolically every summer and had effectively shrunk to the status of a judicial court. It would meet for the last time in 1798 and be abolished two years later.[2] When Sir Joseph Banks' party reached Þingvellir in September 1772 they found only a dilapidated timber building inhabited by what Banks ungraciously called a 'Parson, lazy ignorant as [the] devil'.[3] In fact, Banks and his companions were much

1. See C. M. R. Fowler, *The Solid Earth: an introduction to global geophysics*, 2nd edn (Cambridge, 2005), p.391-93, 585-86.
2. See Anna Agnarsdóttir, 'Introduction: Banks and Iceland', in *Sir Joseph Banks, Iceland and the North Atlantic 1772-1820: journals, letters and documents*, ed. Anna Agnarsdóttir (London, 2016), p.1-34.
3. Sir Joseph Banks, 'The Iceland journal of Sir Joseph Banks, Part II: 17

more interested in the surrounding jagged strata of lava forming the huge fissure in the earth known as the Almannagjá. They measured its uneven sides as rising to 107 feet on the western wall and 45 feet on the eastern. Nor was Banks immune to the beauty of this volcanic landscape, especially to the north where it appeared 'very fine & vastly romantick & pretty', the rocky cleft 'becoming the bed of a river [the Öxará] which falls in Cascades through it'.[4]

Of course, Banks was not the first traveller to visit Iceland in the eighteenth century, and he drew on previous accounts in order to prepare his expedition. As we shall see in this chapter, these bold empirical studies of Iceland in the mid- to late eighteenth century, including those inspired by Banks' explorations of the island, confirm a number of key contemporary tropes of volcanological fieldwork undertaken elsewhere in Europe. They privilege first-hand observations over sailors' tales, indulge in direct ethnographical evaluations of the hardy islanders, affirm the extreme, oxymoronic power of volcanic landscapes, compounded in Iceland by its explosive overlaying of glaciers and geysers, and testify to both volcanic sublimity and a levelling sociability experienced in ranging over miles and miles of eruptive terrain. In addition, the ostensibly disinterested French and British accounts of exploring Iceland in the 1760s and 1770s conflate the geophysical with the geopolitical, subtending their scientific observations with possible colonial designs or colouring them with the language of imperial conquest. In this much, they form a bridge between the old-world association of volcanic might embodied in the spirited Icelandic autonomy declared at Þingvellir and the modern identification of geological upheavals and political revolution that France would seal from 1789 onwards.

Yet, perhaps more importantly, as the second half of this chapter will show, it was specifically the Laki eruption of 1783 and its calamitous global consequences that gave contemporary scientists, state administrators, thinkers and writers the most insistent intimations of the ecological and social implications of violent, large-scale volcanic activity in Europe. In this sense in particular, Laki provided the late eighteenth century with an explosive, apparently remote, yet intensely intimate, ecological corollary of the volcanic sublime, one which, like volcanic sublimity, began to form and inform a modern

---

September-22 October 1772', in *Sir Joseph Banks, Iceland and the North Atlantic*, ed. A. Agnarsdóttir, p.93-114 (95).

4. Banks, 'The Iceland journal of Sir Joseph Banks, Part II', p.93-95.

environmental sensibility. The overly simplistic historical assumptions that the violence of Laki must have somehow precipitated the violence of the French Revolution six years later have been clearly and definitively debunked. Yet it is still true that this twenty-seven-kilometre gash in the Earth's crust, spewing out square hectares of lava and megatons of volatile ash clouds, offered a singularly potent illustration of the interrelation of volcanism and the fragile equilibriums regulating both natural and social ecologies at the time.

However, we will begin with Banks' predecessors and instructors in the volcanic peculiarity of eighteenth-century Iceland. The most famous of these was the Dane Niels Horrebow and his seminal work *Tilforladelige efterrretninger om Island* (1752), translated into German in 1753, Dutch in 1754 and English in 1758 under the title *The Natural history of Iceland.* A jurist and amateur scientist, Horrebow had lived in Iceland from 1749 to 1751 and made copious notes on the natural history of the island and the customs of its inhabitants. His study was meant above all to be a thorough rebuttal of Johann Anderson's *Nachrichten von Island* (1746), a highly coloured compilation of dismissive impressions of Iceland, culled in part from Danish sailors' tales in Hamburg. As far as the island's 'burning mountains' are concerned, Horrebow seeks systematically to refute Anderson's claims that Iceland is constantly riven by seismic concussions and volcanic conflagrations, leaving it little more than a barely inhabitable desert of 'ash and dust of burnt rocks'.[5] In contrast, his own travels over the island revealed fewer than twenty historically active craters.[6] Horrebow is similarly keen to dispel the myth that the most notorious Icelandic volcano, Hekla, flared into life as the result of subterranean communication with the Italian volcanoes of Vesuvius and Etna. This is a vision of globally interconnected volcanism borrowed from Athanasius Kircher, with the Earth traversed internally by tremendous channels of fire, bursting out as volcanoes wherever they breach the surface. To reject this idea, Horrebow points empirically to the notable eruptions of both Vesuvius and Etna at the start of the 1750s while Hekla

5. Johann Anderson, *Nachrichten von Island, Grönland und der Strasse Davis, zum wahren Nutzen der Wissenschaften und der Handlung* (Hamburg, Georg Christian Grund, 1746), p.3-11 (10).

6. Niels Horrebow, *The Natural history of Iceland: containing a particular and accurate account of the different soils, burning mountains, minerals, vegetables, metals, stones, beasts, birds and fishes, together with the disposition, customs, and manner of living of the inhabitants* (London, A. Linde, 1758), p.11.

remained stubbornly quiet.[7] Later authors on Iceland were to reiterate this scepticism regarding the notion of vast underground passages connecting European volcanoes, although the idea may well have inspired Jules Verne's *Voyage au centre de la Terre* (1864) insofar as his intrepid adventurers enter Snæfellsjökull on the western tip of Iceland and are spat out of Stromboli off the coast of Sicily.[8] Instead, Horrebow posits a more localised explanation for volcanic activity, namely, the explosion of substantial stocks of highly flammable sulphur in volcanic vents. This draws on Lémery and Buffon's chemical understanding of volcanism. Yet Horrebow illustrates the theory with the specific historical example of the 1728 eruption of the Krafla fissure in northern Iceland, which 'set the sulphurous earth around in flames, melted and made it run like water'. This was in fact just one of the more fiery episodes in a longer burst of volcanic activity in the Krafla region running from 1724 to 1729. The particular outpouring of lava described by Horrebow consumed three farmsteads and spilled into the Mývatn lake, boiling and poisoning its fish stocks.[9] The Danish author is nonetheless keen to relativise the volcanic dangers in Iceland. He recognises that what is unique to the island is the lethal overlaying of volcanoes with glacial ice caps. Such mountains are called *jökulls* in Icelandic. They threaten inhabitants and travellers alike with the double dangers of spurting fire and swiftly swollen torrents of glacial melt, as in the 1721 eruption of Katla which allegedly flash-thawed thousands of tons of ice and snow, sweeping volcanic debris nearly three miles out to sea.[10] Yet other eruptions were tamer, with Horrebow citing a local man who had visited Krafla in 1728 and found the lava flowing so 'even and quiet' that he 'lighted his pipe by it'.[11] Other dreaded volcanoes proved to be disappointingly dull. An established tradition going back to the Cistercian monks in the twelfth century and revived by the Protestant physician Kaspar Peucer in the fifteenth century held Hekla to be nothing less than the gates of hell from which rose the 'melancholy cries and loud wailings' of damned souls.[12] Yet Horrebow reports that, when the Icelandic students Eggert

7. Horrebow, *The Natural history of Iceland*, p.15.
8. Jules Verne, *Voyage au centre de la Terre* (Paris, 1864).
9. Horrebow, *The Natural history of Iceland*, p.9-10.
10. Horrebow, *The Natural history of Iceland*, p.11-13. The volcano is called 'Kotlegau' in the English text.
11. Horrebow, *The Natural history of Iceland*, p.11.
12. See Bernard de Clairvaux, *De miraculis* (1180), and Kaspar Peucer, *De dimensione terræ* (1550); see also David K. Chester and Angus M. Duncan, 'Geomythology,

Ólafsson and Bjarni Pálsson first conquered the volcano in June 1750, they found nothing of note on the summit and were troubled only by wading through knee-deep screes of ash and cinder.[13]

In fact, Ólafsson and Pálsson would go on to produce the most comprehensive study of their native land written in the eighteenth century. On finishing their studies at the University of Copenhagen, the young conquerors of Hekla were chosen by the Danish Academy of Sciences to carry out a much more detailed exploration of the island. This they did from 1752 to 1757, publishing their findings in two thick tomes in 1772 under the title *Reise igiennem Island.* Their 'journey through Iceland' involved a return to Hekla where they debunked another enduring myth, namely, that its hellish crater was guarded by a flock of iron-beaked crows. Drawing on ancient annals and other local sources, they estimated that the volcano had erupted twenty-two times in recorded history (revising upwards Horrebow's list of ten eruptions, the last occurring in 1693).[14] Their extensive travels took them across varied volcanic landscapes, establishing distinctions between the different types of lava, or *hraun* as it is called in Icelandic, from the verb *hrinna*, meaning 'to flow'.[15] Yet the undoubted volcanic highlight of their expedition was witnessing the massive eruption of Katla which began on the morning of 17 October 1755. The peak, to the east of the Eyjafjallajökull glacier, spewed out an estimated 1.5 cubic kilometres of searing volcanic matter and dissolved its ice cap into a thunderous torrent discharging the same amount of water per second as the world's four largest rivers combined. The ice sheets exploded with terrifying groans and cracks while the surging floodwaters transported, in Ólafsson and Pálsson's words, slabs of ice 'as large as small mountains' and 'boulders the size of houses'.[16] The first eruptive phase lasted a month, after which the

---

theodicy, and the continuing relevance of religious worldviews on responses to volcanic eruptions', in *Living under the shadow: the cultural impacts of volcanic eruptions*, ed. John Grattan and Robin Torrence (Walnut Creek, CA, 2007), p.203-24 (206). The mythical function of Hekla was idiosyncratically revived in the late eighteenth century by William Blake in his *Poetical sketches* (London, n.n., 1783), in which Hekla becomes the icy lair of banished 'Winter' (p.4).

13. Horrebow, *The Natural history of Iceland*, p.16.
14. Eggert Ólafsson and Bjarni Pálsson, *Voyage en Islande*, translated by Gauthier de La Peyronie, 5 vols (Paris, 1802), vol.5, p.28-44; Horrebow, *The Natural history of Iceland*, p.15-16.
15. Ólafsson and Pálsson, *Voyage en Islande*, vol.3, p.383.
16. Ólafsson and Pálsson, *Voyage en Islande*, vol.4, p.264-65.

crater was intermittently active until September 1756. The volcanic explosion and catastrophic flooding destroyed fifty farms, smothering the plains below in two feet of pumice, ash and grit. A lot of livestock was killed or driven mad by the eruption, although amazingly there were only two human fatalities, both bizarrely blasted by volcanic lightning, with one victim flash-fried in clothes that were singed but not burned.[17] Ólafsson and Pálsson made a number of interesting observations during the eruption, recording the unusual glowing 'balls of fire' that the volcano threw out. These sizable pyroclasts or volcanic bombs appear to be a distinctive feature of Icelandic eruptions, since eyewitnesses of the Krafla lava flows of 1728 also told the authors that 'globes of fire' had been frequently hurled from the burning fissure.[18] Ólafsson and Pálsson were also inclined to deduce from the dirty, salty deluge pouring down the volcano that seawater infiltration was the most probable trigger for eruptions.[19] In this they seem to have been misled by the sheer volume of glacial melt caused by Katla's fires, with the meltwaters incorporating large quantities of mineral-rich debris as it swept down the mountainside. Their thinking may equally have been influenced by the prevailing theories of Buffon and his fellow Neptunists.

Hekla remained dormant while Ólafsson and Pálsson were exploring the region; they noted that it too had spewed out both fire and saltwater in the past, although not in the devastating quantities produced by Katla. Salt was commonly found around Hekla, so much so that horse loads of it had been removed for sale. This reinforced the explorers' assumption that the Icelandic volcanoes communicated directly with the sea, even if, as they acknowledged, the salt could have derived instead from substantial deposits of halite or rock salt in the mountain.[20] In April 1766, nine years after the Icelanders completed their travels, Hekla erupted again, but with little evidence this time that it vomited seawater. In fact, the eruption was explosive and sustained, continuing in one form or another until March 1768. As a result, it extruded the largest lava flow produced by the volcano in historical times, covering sixty-five square kilometres.[21] A French expedition

17. Ólafsson and Pálsson, *Voyage en Islande*, vol.4, p 271-75.
18. Ólafsson and Pálsson, *Voyage en Islande*, vol.4, p.266, 268; on Krafla, vol.4, p.211.
19. Ólafsson and Pálsson, *Voyage en Islande*, vol.4, p.279-80.
20. Ólafsson and Pálsson, *Voyage en Islande*, vol.5, p.34-36.
21. See Agnarsdóttir, 'Introduction: Banks and Iceland', p.20.

led by Yves-Joseph de Kerguelen-Trémarec arrived in Iceland in spring 1767. They passed Cape Hekla, as they called it, on 12 May, but their subsequent accounts made no reference to signs of volcanic activity. They eventually anchored in the far north-western harbour of Patreksfjörður on 30 May. Their ostensible mission was to protect the fleet of French fishing vessels, sailing mainly out of Dunkerque, which trawled the Icelandic waters for cod. Hence, as Kerguelen explains in his *Relation d'un voyage dans la mer du Nord* (1771), he and his crew did not explore very far inland; in fact, his only reflection on the recent 1766 eruption of Hekla concerns the unimaginable quantity of tephra that the volcano ejected, allegedly peppering the sea with fuming pumice to a distance of sixty miles from the shore.[22]

If this last figure is somewhat exaggerated, it nonetheless conforms to the disturbing vision that the French sailor had of the highly unstable, volatile Icelandic topography. He writes: 'Peaks were suddenly seen to fall and lakes to form, *jökulls* or ice mountains melted, caught fire, and combined the twin horrors of being engulfed in water and consumed in flame.'[23] He also made unsubstantiated claims that the island contained valuable marble, crystal, copper and iron deposits.[24] He was noticeably more accurate in his charting of the Icelandic waters and coastline, and more generous in his assessment of the inhabitants as 'good, gentle and humane' (even if they might also be 'lazy, distrustful drunkards').[25] Kerguelen cites Horrebow, as do all late-eighteenth-century visitors to Iceland, although he adds the proviso that, as a Dane, Horrebow is wont to minimise the natural disasters ravaging a territory that belonged to the Danish crown.[26]

Overall, then, Kerguelen's description of Iceland is mixed. Yet there might be a more insidious coherence to his hydrography, topography, mineralogy and ethnography if, as Anna Agnarsdóttir suggests, shadowy French colonial designs on Iceland lay behind Kerguelen's apparent mission to escort fishing boats to the north Atlantic. The plan would involve Iceland becoming a French colony in exchange for the Danish crown obtaining a relatively unproductive

22. Yves-Joseph de Kerguelen-Trémarec, *Relation d'un voyage dans la mer du Nord, aux côtes d'Islande, du Groenland, de Ferro, de Schettland, des Orcades & de Norwége, fait en 1767 & 1768* (Paris, Imprimerie de Prault, 1771), p.35.
23. Kerguelen-Trémarec, *Relation d'un voyage dans la mer du Nord*, p.39.
24. Kerguelen-Trémarec, *Relation d'un voyage dans la mer du Nord*, p.39-40.
25. Kerguelen-Trémarec, *Relation d'un voyage dans la mer du Nord*, p.60.
26. Kerguelen-Trémarec, *Relation d'un voyage dans la mer du Nord*, p.39.

expanse of Louisiana. Then from their new naval base in Iceland, the French could launch an offensive to retake eastern Canada from the British.[27] Only limited proof has so far been unearthed to support this theory, but it would explain the sailor's obsession with accurate mapping of the island, his assertion of the docility of its people and his inflated representation of its riches in rock and metal ores.[28] The French were certainly interested enough in Kerguelen's findings to send another expedition to Iceland in 1771 under the captaincy of Jean René Antoine de Verdun de La Crenne. However, this mission proved disappointingly inconclusive. Certainly, from a geological perspective, it added nothing of note to Kerguelen's occasional comments on the island's unique volcanism.[29]

The next, and possibly most celebrated, exploration of Iceland in the eighteenth century also had its roots in colonial ventures, but in a much more tangential fashion. This was Sir Joseph Banks' expedition to Iceland undertaken in the late summer of 1772. Banks had already explored Labrador and Newfoundland in 1766 and, more famously, took part in James Cook's first expedition to the South Seas from 1768 to 1771.[30] Returning laden with botanical specimens and fêted as an enlightened explorer, Banks proposed joining Cook's second voyage to the South Seas in 1772, but with an extended retinue of twenty scientists and servants. However, on inspecting the shipboard facilities provided for his party, Banks judged them grossly inadequate and withdrew from the expedition in a fit of pique. Desperate to keep his crew together and employ them in an alternative, glorious undertaking, Banks hit upon the idea of a voyage of discovery to Iceland. Given its northern latitude and the likelihood of a late summer arrival, botany would not be his principal motivation. Instead, he told

27. Anna Agnarsdóttir, 'Iceland in the eighteenth century: an island outpost of Europe?', *Sjuttonhundratal* (2013), p.11-38 (19-20).
28. Kerguelen used the same exaggeration of colonial spoils to justify a return to the (partly volcanic) islands in the southern Indian Ocean that he had discovered and named after himself in 1772, a fruitless and scandalous second expedition for which he was accused of defrauding the state, overstaying on the Ile de France and having women on board his vessel. He was consequently imprisoned on his return to France in 1774. See Jonathan Lamb, 'Inchoate possession: how Captain Kerguelen claimed an island', *Journal for maritime research* 7:1 (January 2005), p.1-15 (4).
29. Agnarsdóttir, 'Iceland in the eighteenth century', p.21-22.
30. The standard biography is Harold B. Carter, *Sir Joseph Banks 1743-1820* (London, 1988).

the Danish envoy in London, to whom he applied for a passport, that his main scientific incentive in visiting Iceland was 'for the purpose of observing Mount Hekla and the Faroe Islands'.[31] The introduction to Banks' own journal for the expedition confirms this new geological objective, stating that 'the many Volcanoes with which it [Iceland] is said to abound made it very desirable to Explore.'[32] Banks' crew of twenty included three companions who had made the journey to the South Seas with him: Daniel Carlsson Solander, a Swedish botanist and erstwhile student of the great Linnaeus in Uppsala; and two long-time servants James Roberts and Peter Briscoe. Roberts also kept a journal of the Iceland expedition and his notes provide interesting corroboration or elaboration of Banks' own account. The party also comprised Dr James Lind, a Scottish astronomer and physician; the artist and engraver John Cleveley, Jr, along with the draughtsmen brothers James and John Frederick Miller; and, last but not least, Banks' French cook Antoine Douvez. A latecomer to the expedition was the Swedish scholar Uno von Troil. A friend of Solander, von Troil had travelled widely in Europe and had a particular interest in the Icelandic language. Hence when he met Banks in late spring 1772 he was invited to join the crew and would ultimately produce the most complete account of the expedition in 1777.[33]

Banks on Staffa and Hekla

Banks and his party boarded the brig *Sir Lawrence* at Gravesend on 12 July and would reach Iceland on 28 August. However, it could be argued that their most enduring legacy for the history of volcanology was not found in Iceland at all but was encountered en route in the Inner Hebrides of Scotland. On 12 August at Morvern on the isle of Mull Banks met an excited Englishman by the name of Leach who told him of an island, allegedly unknown to the Highlanders, composed of 'pillars like those of the Giant's Causeway' in Northern Ireland. Banks set off immediately for the island and arrived there

31. See Banks to Baron Diede von Fürstenstein, 2 July 1772, in *Sir Joseph Banks, Iceland and the North Atlantic*, ed. A. Agnarsdóttir, p.155.
32. Sir Joseph Banks, 'The Iceland journal of Sir Joseph Banks, Part I: 12 July-6 September 1772', in *Sir Joseph Banks, Iceland and the North Atlantic*, ed. A. Agnarsdóttir, p.45-92 (47).
33. Agnarsdóttir, 'Introduction: Banks and Iceland', p.10; Banks, 'The Iceland journal of Sir Joseph Banks, Part I', p.46, 48.

in darkness. This was the island of Staffa. There, as dawn broke, the explorers were greeted with a unique and wonderful sight. Aligned before them were vast, regular colonnades of dark prismatic basalt – 'a scene of magnificence which exceeded our expectations', as Banks put it in his journal.[34] Indeed, the seasoned traveller's first impressions are nothing short of rapturous and reach for the language of the natural sublime:

> Compard to this what are the Cathedrals or the palaces built by man. Mere models or playthings imitations as diminutive as his works will always be when compard to those of nature. Where is now the boast of the Architect. Regularity, the only part in which he fancied himself to exceed his mistress nature is here found in her possession & here it has been for ages uncounted.[35]

Proceeding then in stunned silence along the shore, they came upon 'the mouth of a Cave the most magnificent I suppose that has Ever been describd by traveler'.[36] This is the volcanic sea cavern, formed of amazingly regular polygonal basalt columns, popularly known as Fingal's Cave, named after the hero of James Macpherson's epic poem published in 1761 and purportedly translated from fragments left by the Celtic bard Ossian. Roberts too marvelled at the striking girdle of pillars around the island, giving a more deistic turn to his sense of awe: 'the whole together must impress a thinking mind with the power and wonderful greatness the immensity, and omnipotence of the Author of Nature'.[37] Banks and his party were nonetheless aware that their travels would be judged as much on their contribution to furthering knowledge as on their exercises in exalted imagination. So they set to describing Staffa's 'productions more Philosophicaly', poring over Fingal's Cave for the next twelve hours or so and measuring both cave and island systematically.[38] Banks makes only a tentative allusion to the volcanic origins of the pillars of rock, identifying them as coarse basalt and comparing them to 'the Surface of a Lava'. Roberts goes further, however, in supposing them 'to have been produced by firey

34. Banks, 'The Iceland journal of Sir Joseph Banks, Part I', p.67-68. Spelling as in the original journal.
35. Banks, 'The Iceland journal of Sir Joseph Banks, Part I', p.68.
36. Banks, 'The Iceland journal of Sir Joseph Banks, Part I', p.68.
37. James Roberts, 'The Iceland journal of James Roberts', in *Sir Joseph Banks, Iceland and the North Atlantic*, ed. A. Agnarsdóttir, p.115-40 (125).
38. Banks, 'The Iceland journal of Sir Joseph Banks, Part I', p.69-72.

Irruptions from the Bowels of the Earth' at an inconceivably remote time in the history of the world.[39]

Neither Banks nor Roberts published their journals. What popularised Banks' enthusiastic descriptions of the volcanic pillars and sea caves on Staffa was their inclusion in the second edition of Thomas Pennant's *A Tour in Scotland, and voyage to the Hebrides 1772*, published in 1774. Pennant, a Welsh zoologist and geologist, was on his second expedition to Scotland at the same time that Banks and his party sailed through the Inner Hebrides. Bad weather prevented Pennant from visiting Staffa himself so Banks, feeling that the Welshman had prior publishing rights on the region, offered him his journal to supplement his writings. He also gave Pennant some important drawings by Cleveley and John Frederick Miller of the stunning basalt topographies on the island. Pennant's book, embellished with Banks' notes and his draughtsmen's illustrations, was hugely popular.[40] It fed a later vogue to visit Staffa as an essential part of any exploration of western Scotland. The French geologist Barthélemy Faujas de Saint-Fond travelled to England in 1784, met Banks in London and eagerly pursued his journey north to see the basalt columns on Staffa. Having championed the volcanic nature of similar prismatic rock features in Auvergne in 1778, Faujas asserted categorically not only that the basalt pillars and cave of Staffa were the products of volcanism, but that 'the extinct volcano of the isle of Staffa', eroded by stormy seas over an incalculable span of time, stands now as 'the skeleton of a once much more considerable volcanic island'.[41] In Faujas's wake Staffa was visited in the nineteenth century by literary celebrities (Walter Scott, Jules Verne, Robert Louis Stevenson), artists (J. M. W. Turner), composers (Felix Mendelssohn), explorers (David Livingstone) and royalty (Queen Victoria and Prince Albert).

If further publicity was required for Banks' enthusiastic description of Staffa, it was to be found in his companion Uno von Troil's *Bref rörande en resa til Island MCCLXXII* (1777), the Swede's extended reflections on their Iceland expedition, translated into German in

39. Banks, 'The Iceland journal of Sir Joseph Banks, Part I', p.71-72; Roberts, 'The Iceland journal of James Roberts', p.125.
40. See Pennant, *A Tour in Scotland*, p.299-309; on Banks and Pennant, see John Gascoigne, *Joseph Banks and the English Enlightenment: useful knowledge and polite culture* (Cambridge, 2003), p.88-98.
41. Faujas de Saint-Fond, *Voyage en Angleterre, en Ecosse et aux îles Hébrides*, vol.2, p.69. On Faujas's Plutonism, see chapter 2 in the present volume.

1779, English in 1780 and French 1781. This work also incorporated Banks' account, only omitting the measurements of Fingal's Cave which von Troil had reproduced elsewhere in his text. Interestingly, he chose to preface Banks' description with his own impressions of the basalt columns of Staffa. Unlike Banks, the Swede had little time for the romanticised tales of Gaelic legend that recounted how a prehistoric race of giants had stacked the pillars with 'supernatural force'. Moreover, he pre-empted Faujas in clearly affirming the volcanic origins of the basalt columns, claiming that they only differ from similar volcanic formations found across Germany, Italy, France, the British Isles and Iceland by their exemplary cleanness of line.[42] But what is perhaps most telling, in terms of the evolution of intellectual history regarding volcanoes, is how von Troil subsumes the rock type in the rock form. That is, his stated object is 'those large regular pillars known by the name of Basalts', whereby basalt is identifiable less by its mineral composition than by the regular, polygonal shapes it takes.[43] It is classified here primarily by its form rather than its formation. Of course, von Troil still draws on the classical Enlightenment device of analogy, echoing both Banks and Roberts in likening a scarp of curved basalt pillars to the 'interior timber-work of a ship'.[44] He also, like Banks, resorts frequently to an architectural vocabulary to define the columnar arrangements of purely geophysical structures. However, possibly under the influence of Cleveley and Miller's drawings, von Troil appears to move from understanding the works of volcanism on Staffa by way of discursive resemblances towards a comprehension of them based on their recurrent visual patterns. This is consonant with the advent of crystallography as a discrete science in the 1770s, apparently confirming the primacy of form in mineralogical and geological analyses over other physical attributes of rocks, such as hardness, density or combustibility.[45]

42. Uno von Troil, *Letters on Iceland: containing observations on the civil, literary, ecclesiastical, and natural history, antiquities, volcanos, basaltes, hot springs, customs, dress, manners of the inhabitants, &c., &c.* (London, W. Richardson, 1780), p.266-87; see Banks' account, p.288-93.
43. Von Troil, *Letters on Iceland*, p.266.
44. Von Troil, *Letters on Iceland*, p.275.
45. See Denise Longeot, 'La genèse de la cristallographie et de la minéralogie scientifique', *Dix-huitième siècle* 3 (1971), p.253-64. One of the seminal works in the field was published the same year as Banks' expedition to Iceland, Jean-Baptiste Romé de L'Isle's *Essai de cristallographie* (1772).

Banks was struck by the sight of a further expanse of basalt when, at the end of August, the *Sir Lawrence* finally came to anchor in the port of Hafnarfjörður in south-west Iceland. However, this was not prismatic basalt as on Staffa, but a large lava field, estimated by von Troil to run for sixty to seventy miles inland, composed of sharp ridges of dark rock. Here Banks felt more assured in asserting that this bleak topography had been 'occasioned by the operations of fire'.[46] However, in the following six weeks or so that he and his crew spent in Iceland, they would be disappointed not to come across any fierily active volcanoes. It was not for want of searching. After two weeks exploring the Hafnarfjörður area, Banks and a party of five set out with nineteen packhorses for a 350-mile round trip to Hekla that would take them twelve days. As we know, they passed through Þingvellir, and its unique union of geology and politics, before coming across their first natural wonder: the boiling waters of Laugarvatn. Yet it was the following day, 21 September 1772, that they saw their first volcano. Amazingly, however, this was not formed of basalt or any other rock but was, in Banks' words, 'A Volcano of Water & stones'.[47] In what was to become a classic tourist destination following in their footsteps, the travellers had arrived at Geysir, an immense spouting hot spring that would subsequently give its name to all the other geysers in the world. Banks and his companions camped by this large cauldron of boiling water, timing its 'eruptions' and measuring water temperatures. They estimated that the geyser was five fathoms (thirty feet) deep and gushed water up to ten fathoms (sixty feet) high.[48] Less scientifically, they shot and boiled ptarmigan in the steaming, jetting waters, finding that the bird was cooked in less than seven minutes. Roberts was fascinated with these northern cousins of the grouse, deeming them 'very silly Birds' for being so tame, yet also 'delicious Eating' – the one quality facilitating the other.[49]

After a further four days of trekking, stopping at the bishop's residence at Skálholt and then fording the river Þjórsá, they eventually reached the foothills of Hekla, entering on a 'Scene of Ashes &

46. Von Troil, *Letters on Iceland*, p.4; Banks, 'The Iceland journal of Sir Joseph Banks, Part I', p.83.
47. Banks, 'The Iceland journal of Sir Joseph Banks, Part II', p.93-109 (98).
48. Banks, 'The Iceland journal of Sir Joseph Banks, Part II', p.99-100. In fact, the largest jets have been measured at seventy metres.
49. Banks, 'The Iceland journal of Sir Joseph Banks, Part II', p.99; Roberts, 'The Iceland journal of James Roberts', p.133-34.

desolation all round almost inconceivable'.[50] It was 4am on 25 September when they began their ascent of the mountain. After two hours on horseback, they came to steep lava fields and proceeded on foot. The climb in itself was not difficult, however the party was beset by a furious, icy wind. Both Roberts and von Troil enlarge on this hardship, claiming that the blasts were so violent that they had to lie down in places to avoid being blown into nearby precipices or craters.[51] When they reached the first peak, their benumbed Icelandic guide tried to convince them that it was the top of the mountain, until the 'Fog and flying Clouds' parted a moment to reveal the true summit above them. Leaving Solander and the guide, the rest of the party 'Hurried up to the top'.[52] There the intrepid climbers huddled victoriously in air temperatures of 24°F (−4°C), lowered much further by ferocious wind chill. They discovered no crater, only a snow-free patch of fumeroles about the size of a tennis court that, according to Banks, hissed with steam and was too hot to 'set [sit] upon'. This last comment would make sense if von Troil's reading is correct that the ground temperature averaged 153°F (67°C).[53] With their Ramsden barometer, they calculated the height of Hekla at 5000 feet (it stands now at 4892 feet). However, given the freezing gale and scudding clouds, there was little else worthy of observation on the summit. So, pulling their ice-matted clothes around them, Banks and his fellow explorers set off briskly for the descent. They scrambled down through snowfields and over the lava produced by the 1766 eruption, reaching their tents again at 2pm, 'a good deal fatigued' but safe.[54]

Scientifically, Banks' conquest of Hekla provided little new information. However, it afforded three separate accounts of the ascent – by Banks, Roberts and von Troil – which are marked by some recurrent cultural themes of eighteenth-century volcanic exploration. For instance, while Hekla's frozen summit might not have revealed the sublime lava-filled chasms of Vesuvius or Etna, it prompted reflection on the intensely oxymoronic experience of being exposed 'at one, and the same time, [to] a great degree of heat and cold, the Mountain

50. Banks, 'The Iceland journal of Sir Joseph Banks, Part II', p.106.
51. Roberts, 'The Iceland journal of James Roberts', p.135; Von Troil, *Letters on Iceland*, p.241.
52. Banks, 'The Iceland journal of Sir Joseph Banks, Part II', p.106-107.
53. Banks, 'The Iceland journal of Sir Joseph Banks, Part II', p.107; Von Troil, *Letters on Iceland*, p.242.
54. Banks, 'The Iceland journal of Sir Joseph Banks, Part II', p.107.

beneath being so hot and the Air above so cold'.[55] Similarly, as on the Italian volcanoes, the practical response to Hekla's icy, thin air was to reach for 'a hearty draught of Brandy'.[56] The liquor was solicitously passed around and Roberts, the servant, was struck by the unusual equality that reigned among the party on the volcano. He recounts how the six travellers drew lots, without favour or preferment, to see who should sleep at the coldest edges of their communal tent. The anecdote, he continues, 'is mentioned to prove how Amicable our little Society was, and that no distinction was made in regard to Superiority in point of Fortune, here was no Master and Slave, but a willingness to assist each other put us all upon an equality, except in point of Learning and Philosophical knowledge'.[57] The volcano's levelling of social hierarchies is accompanied by a reinvigorated collective spirit to defy the dangers of the mountain. Von Troil proclaimed that from the outset 'There was not one in our company who did not wish to have his cloaths a little singed, only for the sake of seeing Heckla in a blaze', while Roberts reaffirmed at the foot of the mountain that even the prospect of a cataclysmic eruption 'had no weight with our Philosophical Adventurer's [*sic*]'.[58] The explorers' hubris here might be read as yet another attempt to arrogate the sublimity of the mountain to those who are prepared to confront and 'conquer' it. Certainly, all three accounts of climbing Hekla glory in the falsehood that Banks' party was the first to reach the summit of the volcano. This claim was also widely trumpeted in the newspapers when the heroic travellers returned to Britain in mid-November 1772.[59] What is surprising, in a sense, is that Banks allowed this erroneous assertion to stand, and might even have encouraged its circulation. For in Iceland he had met one of the first men to conquer Hekla in 1750, Bjarni Pálsson, and it is highly likely that they had touched on their respective ascents of the volcano, even if no written trace of this particular exchange survives.

Fortunately, Banks brought back more than egotistical boasts from his journey to Iceland. He returned with a large number of rare

55. Roberts, 'The Iceland journal of James Roberts', p.135.
56. Banks, 'The Iceland journal of Sir Joseph Banks, Part II', p.106.
57. Roberts, 'The Iceland journal of James Roberts', p.135.
58. Von Troil, *Letters on Iceland*, p.5; Roberts, 'The Iceland journal of James Roberts', p.135.
59. See, for example, Von Troil, *Letters on Iceland*, p.5; on the newspaper reports, see *The Annual register for the year 1772* (London, n.n., 1773), p.116, or *The London magazine* (London, n.n., 1772), p.509, both cited in Agnarsdóttir, 'Introduction: Banks and Iceland', p.23.

Icelandic manuscripts, including a copy of Snorri Sturluson's *Edda*, two sagas and a sixteenth-century Bible printed in Icelandic at Holár. He also brought back examples of Icelandic dress and farming tools, as well as a large ballast of basalt lava collected around Hafnarfjörður. The lava would ultimately go to form rockeries in Kew Gardens and at Chelsea. Volcanic Iceland marked Banks in other ways too. He came back with two Icelandic dogs named Hekla and Geysir, and his visiting card was reprinted to carry an outline of the island with Hekla prominently marked on it spouting a torch-like flame.[60] In fact, for the rest of his long and illustrious career Banks would remain an interested friend and protector of Iceland, even if, in keeping with the colonial politics of the age, this meant proposing in 1810 that Britain annex the island in order to prevent it falling into the hands of the enemy French.[61] He also inspired other Britons to venture north, such as the young aristocrat John Thomas Stanley who retraced Banks' route via Geysir to the summit of Hekla in 1789. In a flourish of youthful vainglory, Stanley went so far as to plant the British flag on the summit.[62] In the early nineteenth century, he was followed to Iceland and its volcanoes by other explorers aware of Banks' exploits. In fact, in 1809 Banks, by then the renowned president of the Royal Society, sent his protégé William Jackson Hooker to Iceland in order to collect the island's flora, lending the young explorer his own account of the 1772 ascent of Hekla in order to guide him on the volcano.[63] At least as far as Hekla is concerned, we might then trace a shift in the cultural significance borne by this iconic volcano towards the end of the eighteenth century. From being a metonym of all Iceland's terrible volcanism, it became a metaphor of self-interested Icelandic exploration, and more specifically the mark of scientific and quasi-imperial 'conquest'.[64]

60. See Agnarsdóttir, 'Introduction: Banks and Iceland', p.20-22, 24.
61. Anna Agnarsdóttir, 'Iceland under British protection during the Napoleonic wars', in *Scandinavia in the age of revolution: Nordic political cultures, 1740-1820*, ed. Pasi Ihalainen *et al.* (Farnham, 2011), p.255-66; Agnarsdóttir, 'Introduction: Banks and Iceland', p.29-34.
62. See Sir John Thomas Stanley, *The Journals of the Stanley expedition to the Faroe Islands and Iceland in 1789*, ed. John West, 3 vols (Torshavn, 1970-1976), vol.1, p.208. Cited in Agnarsdóttir, 'Iceland in the eighteenth century', p.23-24.
63. See Sir William Jackson Hooker, *Journal of a tour in Iceland, in the summer of 1809*, 2nd edn, 2 vols (London, 1813), vol.2, p.105-19.
64. In this regard it is perhaps instructive that Stanley had already climbed Vesuvius and Etna on his Grand Tour of 1787 and was urged to undertake his

Banks was not the only one of his party to return from Iceland with volcanic souvenirs. On his return to Sweden in 1773 von Troil accompanied one of his letters to his friend, the Swedish chemist and mineralogist Torbern Bergman, with samples taken from his collection of Icelandic lava. Pronouncing himself broadly agnostic on the competing merits of Plutonist and Neptunist theories of the origins of the Earth, von Troil was nonetheless confident that the basalts of Staffa were 'the remnants of an antient volcano' and that Iceland had been 'formed by eruptions of fire'.[65] Bergman was fascinated by the lava samples that he received from his friend. In June 1776, after long reflection on these products of Icelandic volcanism, the chemist sent a fulsome reply to von Troil. While Bergman was on the whole a proponent of Neptunism, favouring water-based explanations for geological phenomena, he conceded to his friend that fire was evidently an active agent in the formation of basalt pillars and Icelandic terrain. He just was not sure how these volcanic processes operated in practice. Consequently, Bergman hedged his bets and generalised that fire was on the whole a more powerful agent in shaping the surface of the Earth, but that its effects were more localised than those of water.[66] Only seven years after Bergman drew these cautious conclusions from the Icelandic lavas sent to him by von Troil, a devastating eruption occurred in southern Iceland whose lethal global fallout proved beyond doubt that large-scale volcanism was rarely, if ever, just a local matter.

The deadliest volcano: Laki 1783

Let us start with the facts as earth and climate sciences currently understand them. On 1 June 1783 a series of strong earth tremors were felt in southern Iceland. On 8 June a twenty-seven-kilometre-long volcanic fissure suddenly tore open around a mountain called Laki in the same region. The fissure was formed of ten vast gashes in the Earth's crust, bursting with more than 140 lava vents, some, spurting fire fountains one to two kilometres into the air. This was

Iceland expedition by Mark Beaufoy, the first Englishman to 'conquer' Mont Blanc in 1787. See Andrew Wawn, 'John Thomas Stanley and Iceland: the sense and sensibility of an eighteenth-century explorer', *Scandinavian studies* 53:1 (winter 1981), p.52-76.

65. Von Troil, *Letters on Iceland*, p.16, 222.

66. Torbern Bergman to Uno von Troil, 'Of the effects of fire, both at the volcanos and the hot springs, and also of the basalts', 12 June 1776, Stockholm, in von Troil, *Letters on Iceland*, p.338-400 (366-67).

the shattering start of the so-called Laki eruption, or more properly Lakagígar, meaning 'Craters of Laki', a series of ten eruptive episodes which would gush with lava for the next eight months. The huge fissure ejected 14.7 cubic kilometres of sulphur-rich basaltic lava, covering 565 square kilometres of land up to a depth of 20 metres on average and 150 metres in some river gorges. It remains the largest lava-flow eruption of the last millennium; only the cataclysmic explosion of nearby Katla in 934 is thought to have spewed more lava as it gouged out the largest volcanic canyon in the world, the Eldgjá.[67]

The Laki eruption produced towering columns of gas, ash, cinder and debris, firing them to altitudes of nine to thirteen kilometres. Its volcanic plume thus reached into the upper troposphere and the lower stratosphere. It is calculated to have injected 122 megatons of sulphur dioxide (SO2) into the atmosphere, as well as 349 megatons of carbon dioxide (CO2) and 235 megatons of water vapour. These toxic gases then combined with atmospheric moisture to form 200 megatons of sulphuric acid aerosols. The huge eruption also unleashed an estimated fifteen megatons of hydrogen fluoride (HF) and seven megatons of hydrogen chloride (HCl), poisonous gases that would be absorbed by plant life and breathed in by wildlife with fatal consequences. The huge quantities of caustic aerosols in the stratosphere and upper troposphere were then transported by the polar jet stream and by prevailing weather fronts in a broad south-easterly direction, forming a continent-wide veil of sickly sulphuric haze, a so-called dry fog. In the short term the effects of this extensive shield of volcanic gases were a drastic increase in humidity and air temperature as well as the triggering of violent electrical storms. The dry fog effectively absorbed infra-red radiation from the sun and heated the ground below. At the

67. In this and the following analyses I draw on the well-established volcanological and climatological research on Laki, specifically John Grattan and Mark Brayshay, 'An amazing and portentous summer: environmental and social responses in Britain to the 1783 eruption of an Iceland volcano', *The Geographical journal* 161:2 (1995), p.125-34; Richard B. Stothers, 'The great dry fog of 1783', *Climate change* 32:1 (1996), p.79-89; Richard Hamblyn, *The Invention of clouds: how an amateur meteorologist forged the language of the skies* (London, 2001), p.46-56; Thorvaldur Thordarson and Stephen Self, 'Atmospheric and environmental effects of the 1783-1784 Laki eruption: a review and reassessment', *Journal of geophysical research: atmospheres* 108 (2003), p.1-29; John Grattan *et al.*, 'Volcanic air pollution and mortality in France 1783-1784', *Comptes rendus geoscience* 337 (2005), p.641-51; Karen Oslund, *Iceland imagined: nature, culture, and storytelling in the North Atlantic* (Seattle, 2011), p.34-48; Oppenheimer, *Eruptions that shook the world*, p.269-94.

same time it acted as a blanket of greenhouse gases and trapped the heat reflected back from the Earth's surface, thereby exacerbating the ambient levels of heat and humidity.[68] As a result, in western Europe the month of July 1783 was particularly torrid, the stifling oppression broken only by devastating lightning blasts and thunderstorms. The sultry, still air and increased wetness from the flash storms also bred disease. The impact was just as terrible further afield. The toxic volcanic haze tempered the western jet stream, lowering the temperature differential between Atlantic and Indian Oceans, thereby severely weakening the African and Indian monsoon circulation. This led to a reduction in rainfall over eastern Africa, with the annual life-giving Nile flood diminished by 18 per cent, the lowest on record for the period between 1737 and 1800. Contemporary observers speak of a sixth of the Egyptian population wiped out over the following two years as a result of drought, famine and disease, including plague. In India, it is thought that weaker monsoons resulting from the eruption's smothering effects contributed to the dreadful 'Chalisa' famine that amassed a death toll in the millions.[69]

While they were on a smaller scale, the consequences across Europe for livestock and human populations proved no less deadly. By early 1784 in Iceland 80 per cent of the sheep, 75 per cent of the cattle and 50 per cent of the horses had perished. The human population fared little better, decimated by famine and fluoride poisoning. It is estimated that by 1786 the Icelandic population had been reduced by about 20 per cent from 49,000 before the eruption to 39,000 in its desolate aftermath.[70] This was little more than the island's population had been at the institution of the Alþingi in the tenth century. In England the late summer of 1783 saw an unprecedented spike of 30,000 more deaths than average, at a time of year usually considered a low point in mortality rates for the period. France suffered at least the same increase in unseasonal deaths, with the toll particularly heavy among the newborn, where mortality doubled. In fact, across large swathes of western Europe in the summer of 1783,

68. Emmanuel Le Roy Ladurie has claimed that the Laki eruption recreated, on a small scale, the climatic conditions that led to the cataclysmic Permian–Triassic extinction event approximately 252 million years ago. See his *Histoire humaine et comparée du climat II: disettes et révolutions 1740-1860* (Paris, 2006), p.111-21 (119).

69. See Stothers, 'Volcanic dry fogs', p.713-23; Oppenheimer, *Eruptions that shook the world*, p.292-93; and Alan Mikhail, 'Ottoman Iceland: a climate history', *Environment history* 20 (2015), p.262-84.

70. Agnarsdóttir, 'Introduction: Banks and Iceland', p.17.

outbreaks of lethal endemic diseases – malaria, smallpox, typhus, dysentery, influenza – were more frequent and persistent than usual, often compounded by acute respiratory afflictions aggravated by the pervasive air pollution. Such diseases were especially rife among the sick and elderly, further enfeebled by dehydration in the ongoing heatwave. Among the young, it was diarrhoea that often proved fatal, symptomatic of a hot-house climate fostering insect swarms, rotting foodstuffs and tainted water supplies.[71]

As September 1783 wore on, however, across the northern hemisphere temperatures started to drop and the air began to clear. Disturbingly, though, temperatures did not stop falling and the already embattled populations of western Europe braced themselves for the second phase of Laki's brutal and deadly climate disorder. In the medium term, as winter approached, the veil of sulphuric aerosols now reversed the greenhouse-gas effect of the summer. The persistent volcanic haze reflected solar radiation back into space and trapped chilling air beneath, accelerating the winter cooling process. This would continue until at least March 1784. It resulted in an average temperature decrease of 1.5°C in western Europe over the period from 1784 to 1786, with its severest effects felt in eastern North America. Here the temperatures in the winter of 1783-1784 in New Brunswick were the lowest in 250 years.[72] While average falls in temperature may sometimes be a crude mechanism for measuring the climate impact of a singular volcanic event, such as the Laki eruption, it is more difficult to discount the volcano's extreme local effects over the course of these winter months. Thus the early onset of an intense cold spell meant that air temperatures in Iceland rarely got above −15°C in the winter of 1783-1784. Across most of northern-western Europe major waterways froze over. The Clyde in Scotland become impassable because of ice floes; the Thames was frozen so thick that carts drove across it; and the Danube in Vienna was solid ice, limiting food and firewood deliveries. In Paris, where the Seine was also an ice-rink, the temperature drop from mid-December 1783 to mid-January 1784 was a stunning 3.4°C compared to the previous winter. To make matters worse, milder temperatures produced a rapid thawing of the ice-clogged rivers

71. See John Grattan, Sabina Michnowicz and Robin Rabartin, 'The long shadow: understanding the influence of the Laki fissure eruption on human mortality in Europe', in *Living under the shadow*, ed. J. Grattan and R. Torrence, p.153-74.

72. Mikhail, 'Ottoman Iceland', p.270; Agnarsdóttir, 'Introduction: Banks and Iceland', p.17.

which, combined with heavy rains, caused large-scale flooding. The first days of January saw a snap thaw in snowy Paris which flooded the streets, drowning a number of unwary residents and their horses. In February 1784 Vienna, Bratislava and Prague were all swamped, while in Dresden the Elbe rose 8.5 metres, doing enormous damage to river-level properties.[73]

The accumulative evidence of these weather events in the wake of the Laki eruption makes a strong case for its devastating global climate impact, not to mention the appalling lot of deprivation and death that it heaped on the late-eighteenth-century populations of at least three continents. Nonetheless, counter-arguments to this apparently evident 'Laki effect' do exist. Thus it has been claimed that this series of climate disasters is potentially attributable not to sustained volcanic pollution, but to highly infrequent but perfectly natural global climate fluctuations. Specifically, the terrible events of 1783 and 1784 might fit with an unusual coincidence of a negative North Atlantic Oscillation and an El Niño Southern Oscillation, that is, a weakening of the largely stable Icelandic Low and Azores High pressure systems combined with a strengthening of air pressure and sea-surface temperatures in the tropics. Proponents of this explanation, such as Rosanne D'Arrigo, thus compare the Laki winter to the abnormally cold winter of 2009-2010.[74] However, it is fair to say that this is a minority position among historians of earth and climate science. It is perhaps more reasonable to suggest that some of the climatic fluctuations described in this severe cooling scenario could have aggravated the climate impact of Laki, rather than superseded it as a primary cause of the widespread hardship and death experienced in 1783-1784.

It is similarly important to point out that neither the volcano nor the climate acted in a social and political vacuum. Poor governance was a significant local factor in intensifying the cruel and often fatal consequences of the Laki eruption. For instance, the volcanic terrain of Iceland hosted a society based principally on meagre subsistence farming, one which consequently lived on the edge of season-to-season

73. Oppenheimer, *Eruptions that shook the world*, p.279-82.

74. See Rosanne D'Arrigo *et al.*, 'The anomalous winter of 1783-1784: was the Laki eruption or an analog of the 2009-2010 winter to blame?', *Geophysical research letters* 38 (2011), L05706, doi:10.1029/2011GL046696 (last accessed 17 December 2018); for a counter-argument, see Anja Schmidt *et al.*, 'Climatic impact of the long-lasting 1783 Laki eruption: inapplicability of mass-independent sulfur isotopic composition measurements', *Journal of geophysical research* 117 (2012), D23116, doi:10.1029/2012JD018414 (last accessed 17 December 2018).

sustainability. Major natural disasters – like the Laki eruption or the smallpox epidemic of 1709 – ravaged an already vulnerable population, and exposed further the belated, ineffectual responses coming from its Danish overlords.[75] In contrast, the French king Louis XVI won popular acclaim in February and March 1784 for distributing charity and compensation to the sum of 3 million pounds (admittedly, a miniscule 1 per cent of treasury revenues) to victims of the heavy snows and devastating flooding in Paris.[76] To varying degrees, the climate chaos unleashed by Laki only made manifest pre-existing social problems, which were then alleviated, or more frequently exacerbated, by the reactions of European polities. It is thus tricky to determine the eruption's effective role regarding longer-term political and historical causality. This can be wildly exaggerated, as in the groundless claims that Laki precipitated the French Revolution six years later.[77] More sophisticated historical analyses can also overstate their case, as when Alan Mikhail pithily sums up the political disintegration of Turkish rule in Egypt after the severely reduced Nile floods of 1783 and 1784. 'In the final analysis', he concludes, 'the Ottomans lost to the volcano.'[78]

Lived experiences of the Laki eruption and its effects 1783-1784

However abstract the scientific and historical overviews might appear, in the aftermath of the eruption and into the following year the daily suffering of many Europeans was raw and real. It is instructive to examine how the fallout from Laki was experienced 'on the ground', so to speak, in order not to lose sight of the human and cultural

75. Oslund, *Iceland imagined*, p.45-48; Oppenheimer, *Eruptions that shook the world*, p.288-89.
76. Oppenheimer, *Eruptions that shook the world*, p.282. Louis-Sébastien Mercier, an eyewitness to the extreme winter hardships of 1783-1784 in Paris, writes that the grateful populace of the city built a 'snow pyramid' on the corner of the rue Coq-Saint-Honoré in honour of the king's munificence. See his *Tableau de Paris*, 2 vols (Paris, 1994), vol.2, p.1366.
77. See Oppenheimer, *Eruptions that shook the world*, p.293-94; for a more detailed refutation, see Emmanuel Le Roy Ladurie's historiography of the climate in France throughout the 1780s, including what he calls the 'Seven seasons of the "trigger"' of Revolution from 1787 to 1789, in Le Roy Ladurie, *Histoire humaine et comparée du climat II*, p.77-180.
78. Mikhail, 'Ottoman Iceland', p.276.

dimensions of this exceptional volcanic event. And, in this sense, nothing is more poignant or inspiring than the relation of the eruption left by the Icelandic pastor Jón Steingrímsson, the so-called Fire-Priest ('eldklerkur'). Steingrímsson wrote two accounts of the Laki eruption, a more factual essay, the *Eldritið* (1788), and the presentation of the same events in his autobiography *Ævisaga séra Jóns Steingrímssonar* (1791), which interprets the disaster and its aftermath more personally and spiritually.[79] According to this last text, the Lutheran priest moved to the parish of Kirkjubæjarklaustur in 1778, finding a happy and prosperous living some twenty miles south of Laki. He had his own livestock and tended to the spiritual needs of 450 parishioners.[80] The year 1783 started like any other and by June the grass was growing 'green and luscious' on the pastures around the church and its scattering of neighbouring farmsteads.[81] Even so, earth tremors had been felt as early as mid-May, and by 1 June they had become so strong that many locals evacuated their homes and slept outside in tents for fear of buildings collapsing. For Steingrímsson, the first manifestation of the volcanic calamity that would befall him and his congregation occurred 'around midmorn on Whitsun, June 8th of 1783, in clear and calm weather' when 'a black haze of sand appeared to the north of the mountains nearest the farms of the Siða area'. By noon there was 'darkness indoors', with the swelling ash cloud billowing black dust and finely spun volcanic glass, called Pele's hair, so that it choked the air and lay thick on the ground.[82] Then rains came and the ash fell like ink. That night the priest saw jetting pillars of fire on the northern horizon.

The next day came the first reports of the nearby river Skaftá drying up. Its rocky gorge then filled with a gushing, glowing torrent of lava which, according to Steingrímsson, 'poured forth with frightening speed, crashing, roaring, thundering. When the molten lava ran into wetlands or streams of water the explosions were as loud as if many cannon were fired at once.'[83] The impact on animal and human

79. The *Eldritið* has been published in English as Jón Steingrímsson, *Fires of the Earth, the Laki eruption 1783-1784*, translated by Keneva Kunz (Reykjavík, 1998); the *Ævisaga séra Jóns Steingrímssonar* has been published as Jón Steingrímsson, *A Very present help in trouble: the autobiography of the Fire-Priest*, translated by Michael Fell (New York, 2002).
80. Steingrímsson, *A Very present help in trouble*, p.176-78.
81. Steingrímsson, *Fires of the Earth*, p.28.
82. Steingrímsson, *Fires of the Earth*, p.25.
83. Steingrímsson, *Fires of the Earth*, p.26.

life was just as dramatic. 'The snouts, nostrils, and feet of livestock grazing or walking on the grass turned bright yellow and raw' as they chafed desperately at the sulphurous ash and spindly cinders to get at the dying plants beneath.[84] Hay was poisoned and burned with a blue flame; water was contaminated, becoming barely drinkable and leaving a foul aftertaste. The air was similarly unbreathable, gagging and sulphurous, so that those with weak chests 'could no more than half-fill their lungs'.[85] Under almost incessant suffocating tephra-fall, the effects of fluoride poisoning were soon evident. Horses lost their hair and flesh, became too weak to work their jaws and were afflicted with 'hard, swollen lumps' on their joints. The bones of sheep softened and bowed, with large growths appearing on their ribs. Cattle too sprouted cysts and their hooves rotted. The priest noted the same disfigurements in his starved parishioners, covered in sores and bristly growths on joints, hands and feet while 'their bodies became bloated, the insides of their mouths and their gums swelled and cracked, causing excruciating pains and toothaches'.[86] These last symptoms also suggest the increasing presence of scurvy in the population.

Throughout the last weeks of June and the beginning of July the region was also hit with jolting earth tremors while a series of fiery surges brought the lava flows ever nearer to Kirkjubæjarklaustur. The eruption seemed to intensify on 18 July, with renewed ash falls which Steingrímsson describes as 'a sandy and muddy ash' that 'dispersed over the ground here so that it blackened'.[87] Two days later, with the river of molten lava advancing upon them, the priest sorrowfully led his wretched congregation into the small church for what he believed would be his last service. As he described it, the scene was infernal:

> As we approached, the clouds of hot vapors and fog coming from the fire farther down the river channel were so thick that the church could hardly be seen [...]. Claps of thunder were followed by such

84. Steingrímsson, *Fires of the Earth*, p.41.
85. Steingrímsson, *Fires of the Earth*, p.41.
86. Steingrímsson, *Fires of the Earth*, p.77; see also Oppenheimer, *Eruptions that shook the world*, p.286-87.
87. This description is taken from a further account of the eruption written by Steingrímsson with S. Ólafsson entitled 'A simple, but true narrative of the eruption in Skaftafellshire in the year 1783', translated and cited in Thorvaldur Thordarson *et al.*, 'The 1783-1785 A.D. Laki–Grimsvötn eruptions II: appraisal based on comtemporary accounts', *Jökull* 53 (2003), p.11-48 (42).

> great flashes of lightning, in series after series, that they lit up the inside of the church and the bells echoed the sound, while the earth tremors continued unabated.[88]

It was in this apocalyptic moment, with the tide of glowing lava looming, that Steingrímsson gave his celebrated 'Fire Mass' (*eldmessa*). He made his sermon slightly longer than usual but saw not a single person stir to flee the advancing volcanic fires. In fact, he claims that everyone was 'completely unafraid', 'each and every person was without fear, asking [God's] mercy and submitting to His will.' After the service, Steingrímsson and his parishioners emerged to find that the lava 'had not come a foot nearer than before'. Instead, 'it had piled up in the same place, layer upon layer, in a downward-sloping channel some 70 fathoms [385 feet] wide and 20 [110 feet] deep, and will rest there in plain sight until the end of the world, unless transformed once again.' What made the congregation's deliverance seem all the more miraculous was that at this very point the tributary rivers of the Skaftá – the Holtsá and the Fjaðará – burst in torrents over the dams of smoking lava and extinguished their fires. 'There was so much water', writes the priest, 'that horses could not cross the river at all by the cloister' for the rest of that day.[89] The seemingly wondrous success of Steingrímsson's humble intercession at Kirkjubæjarklaustur on 20 July 1783 has rightly gone down in the annals of Icelandic folklore. Michael Fell suggests that many consider it 'the most intense collective "faith experience" recorded in Icelandic church history'.[90]

Yet, if it is sometimes claimed as a miracle, it is one that violated no natural law in diverting the lava away from the church. The lava appears to have stalled on the slopes near the church and spilled eastwards, taking its burning havoc onto the coastal plains. Certainly, Steingrímsson sees the Fire Mass less as a unique supernatural intervention than as part of a more general theodicy or divine justice at work in the world. It is of a piece with his belief that the Laki eruption was 'the Lord's chastisements' for the sinfulness of the local population, confirmed in his eyes by two farmsteads consumed in the volcanic flames being those in which 'the fire of dissension […] had burned more hotly' than elsewhere in his parish.[91] Thus the divine

88. Steingrímsson, *Fires of the Earth*, p.48-49.
89. Steingrímsson, *Fires of the Earth*, p.49-50.
90. Steingrímsson, *A Very present help in trouble*, p.380.
91. Steingrímsson, *A Very present help in trouble*, p.180, 183.

punishment meted out to the wicked Icelanders in the form of a volcanic catastrophe, which began on Whitsuntide, forms a Lutheran counterpart to the Catholic interpretations of the Lisbon earthquake, which struck on All Saints' Day, as the expression of God's wrath at the locals' wanton impiety.

The Fire Mass was a blessed momentary respite in a longer series of divine ordeals visited upon the inhabitants of southern Iceland through the Laki eruption. Moreover, the Lord's anger seemed to redouble a week or so after the Fire Mass when, on 29 July, a further immense volcanic fissure erupted to the north-east of Laki, sending a second scourging trail of lava down the dried-up gorge of the river Hverfisfljót. The farmers, peasants and beggars, caught now between two roaring torrents of fire, fled south in terror. Steingrímsson reports hosting each night seven or more terrified and bedraggled strangers fleeing towards the coast.[92] Starvation, disease and accident overtook many who stayed behind. The Kirkjubæjarklaustur pastor buried seventy-six of his parishioners, stocking the corpses in the church, then burying them in mass graves of eight or ten bodies at a time, working with little help under a volcanic hail.[93] Little wonder then that on Christmas Eve 1783, seeing a colossal volcanic cloud bulk and bulge in the northern sky, Steingrímsson recognised not just the effusion of 'a collection of the various mineral vapours arising from the new lava, which practically surrounded us', but also a 'portent', a divine warning of 'the famine and death which was to follow'.[94] In early 1784, as winter bit, he tended to the sick and starving, claiming that for six full weeks 'I did not take my clothes off and hardly stopped day and night.'[95] Yet, although physically broken by the travails that fell to him in the devastating aftermath of the Laki eruption, Steingrímsson survived what became known as the *Móðuharðindi*, the 'Mist Hardships' or 'Famine of the Haze', which claimed the lives of some 10,000 of his fellow Icelanders. What is more, he survived with his Lutheran faith intact, as his two invaluable accounts of the eruption make clear.

Beyond Iceland the lived experience of the Laki eruption was more diffuse, less urgent but no less deadly. The sulphuric veil issuing from the volcano reached northern Scotland as early as 10 June 1783 and

92. Steingrímsson, *A Very present help in trouble*, p.182.
93. Steingrímsson, *A Very present help in trouble*, p.183.
94. Steingrímsson, *Fires of the Earth*, p.64.
95. Steingrímsson, *A Very present help in trouble*, p.184.

spread over the northern hemisphere in the course of the following three weeks. In Caithness, northern Scotland, ash falls and acid rain damaged crops and poisoned livestock, so that 1783 was referred to balefully as 'the year of the ashie'.[96] Further south and across Europe, the first signs of perturbing changes in the climate were atmospheric. The attentive naturalist and inveterate chronicler of local weather Gilbert White of Selbourne, Hampshire, is often quoted on the bizarre climatic conditions of summer 1783. His journal captures brilliantly the strange occurrences in the weather that persisted troublingly from 23 June to 20 July in what he calls an 'amazing and portentous' summer, 'full of horrible phænomena'.[97] It is worth citing him at length:

> The sun, at noon, looked as blank as a clouded moon, and shed a rust-coloured, ferruginous light on the ground, and floors of rooms; but was particularly lurid and blood-coloured at rising and setting. All the time the heat was so intense that butchers' meat could hardly be eaten on the day after it was killed; and the flies swarmed so in the lanes and hedges that they rendered the horses half frantic, and riding irksome. The country people began to look with a superstitious awe at the red, louring aspect of the sun; and indeed there was reason for the most enlightened person to be apprehensive.[98]

While the dim and dreadful appearance of the sun brought descriptions from Milton's *Paradise lost* to the mind of the Hampshire parson, scientific observers in France and Switzerland tried to fix the sun's altered colour as objectively as possible. It was thus a 'yellowish disk', 'pale and whitish', at other times a 'blood-red globe' for Pierre Bertholon; Horace-Bénédict de Saussure ranging through the Alps saw a 'bluish light' emanating from a 'dark red disk', a light source that 'the eyes could fix upon without fatigue'. The amateur meteorologist Louis Cotte, resident at Laon in northern France, described the dimmed sun as glowing with a 'pale orangey light'. Like Gilbert White, Cotte also remarked on the 'sinister ideas taking hold of the people's mind', inducing a terrified panic at the persistent clouding of the sun.[99]

96. Hamblyn, *The Invention of clouds*, p.56.
97. Gilbert White, *The Natural history and antiquities of Selbourne, in the county of Southampton: with engravings, and an appendix* (London, T. Bensley, 1789), p.301.
98. White, *The Natural history and antiquities of Selbourne*, p.302.
99. Bertholon, *De l'électricité des météores*, vol.2, p.129; Saussure, *Voyages dans les Alpes*,

At the same time the volcanic haze troubled other senses. In late June 1783 the physics professor Jan Hendrik van Swinden, living in Friesland in the northern Netherlands, reported a 'sulphurous odour very readily perceived by the senses, crawling through everything, even closed houses'.[100] It brought on coughing fits, wilted plants and corroded brass pillars. The stink of sulphur was perceptible as far south as Naples where the 'bituminous smell' was 'unbearable in certain places'.[101] In the same region, vines and olive trees were parched and browned by the acidic air. Similar stories of the withering effects of a sticky, caustic dew were recorded as far apart as Ipswich in eastern England and Narbonne in southern France.[102] But these insidious, deleterious influences of the Laki dry fogs seemed slight in comparison to the violent electrical storms that the eruption unleashed. These were particularly ferocious in the first two weeks of July 1783. On the night of 10-11 July Saussure was forced to hole up in the seedy Alpine hostel at Grimsel to avoid the thunder and lightning strikes that broke out 'almost without interruption'.[103] The same storm system seems to have battened on Provence a couple of days later.[104] Meanwhile in England local newspapers reported no fewer than twenty-four deaths and twenty-seven serious injuries resulting from the furious thunderstorms of early July. Electrical gales breaking out in disparate parts of the British Isles on 10 July accounted alone for the death of more than eighty livestock, mainly sheep. As with the zigzagging *ferilli* of Katla in 1755, lightning bolts struck and killed two children in Scotland, with the *Edinburgh advertiser* of 15 July 1783 recounting how their limp bodies 'soon grew black' as though burned from the inside out.[105] In continental Europe the ill-advised practice of ringing church bells to warn of imminent storms proved fatal to six bell-ringers in the steeple at Dobraken near Pilsen in Bohemia; another four were badly scorched by the lightning

vol.3, p.440; Louis Cotte, 'Mémoire: sur les brouillards extraordinaires des mois de juin & de juillet 1783', *Journal de physique, de chimie, d'histoire naturelle et des arts* 23 (September 1783), p.202-205 (202-203).

100. Cited in Oppenheimer, *Eruptions that shook the world*, p.278.

101. Michele Torcia, 'Lettre sur le tremblement de terre de Calabre', *Le Journal des sçavans* (December 1783), p.839-41 (841).

102. Grattan and Brayshay, 'An amazing and portentous summer', p.129; Bertholon, *De l'électricité des météores*, vol.2, p.132.

103. Saussure, *Voyages dans les Alpes*, vol.3, p.470.

104. See Bertholon, *De l'électricité des météores*, vol.2, p.142.

105. Grattan and Brayshay, 'An amazing and portentous summer', p.131.

blasts.[106] Fierce thunderstorms continued to break out under the sulphuric veil. As late as 10 September lightning ravaged the Béziers region in the south of France, ripping off weathercocks, electrocuting livestock and claiming more than sixty human lives.[107]

Many deaths caused by the volcanic pollution were less dramatic. They resulted from aggravated respiratory complaints, heart failures and endemic water-borne diseases. A lot of specific illnesses went under the simple catch-all diagnosis of fever. Thus in Bath, England, the local newspaper of 3 July reported that a 'fever of a putrid kind rages in many parts, which the people term the black fever, and carries off great numbers'. In France Cotte noted that 'red fevers and other eruptive diseases are very common', whose general symptoms included 'gangrenous sore throats, and malignant and putrid fevers'. As far south as Catalonia, according to the diarist Rafael de Amat y de Cortada, 'malignant fevers' raged, 'from which no few persons died'.[108] The future pioneering psychiatrist Joseph Daquin wrote in his *Essai météorologique sur la véritable influence des astres, des saisons et des temps* (1784) that noxious air caused an 'epidemic of continuous, intermittent and remittent bilious fevers' throughout the summer and autumn of 1783.[109] As John Grattan and his fellow researchers have shown, contemporary social and religious testimonies corroborate these claims. Thus parish registers in northern-central France record parishioners 'feeling suffocated by a pestilential illness' which in some parts sent 'a third of the population to the grave'.[110] (They also note bitterly that those whom the toxic volcanic air did not kill were often finished off by incompetent doctors.) The poet William Cowper was bedridden with a 'Cold and feverish complaint' in August 1783, but he was sufficiently recovered by early September to observe the much worse ravages of fever among the farming population of Bedfordshire. 'Such multitudes are indisposed by fevers in this country', he wrote, 'that the farmers have with difficulty gather'd in their harvest, the

106. Alwyn Scarth, *Vulcan's fury: man against the volcano* (New Haven, CT, 2001), p.118.

107. Bertholon, *De l'électricité des météores*, vol.2, p.144.

108. See the *Bath chronicle* of 3 July 1783, cited in Hamblyn, *The Invention of clouds*, p.48; Louis Cotte, 'Extrait des observations météorologiques faites à Laon, par ordre du roi, pendant le mois de juillet 1783', *Le Journal des sçavans* (1784), p.48-52 (50-52); for Cortada, see Grattan, Michnowicz and Rabartin, 'The long shadow', p.169.

109. Joseph Daquin, *Essai météorologique sur la veritable influence des astres, des saisons et des temps* (Chambéry, Gorrin, 1784), p.29.

110. Grattan *et al.*, 'Volcanic air pollution and mortality in France', p.647.

laborers having been almost every day carried out of the field incapable of work, and many die.' The following day, he shared the popular fear with another correspondent that this 'Epidemic' would be 'nearly as fatal as a *plague*'. The poet, often affected by bouts of black depression, saw that the situation would only become 'more mortal as the Autumn comes on'.[111]

By January 1784, sunk further in the pits of despair, Cowper wrote to John Newton that 'the weather is an exact emblem of my mind in its present state. A thick fog invelops every thing, and at the same time it freezes intensely.'[112] Yet it was not just neurasthenics like Cowper who were suffering from the Siberian cold in the winter of 1783-1784 sustained by Laki's continuing eruption. Gilbert White of Selbourne reported how the intense cold of December 1783 caused frostbite in men hunting hares or threshing hay in a barn, while in January 1784 water froze solid under people's beds. In London at the end of the same month he watched the mercury in a Fahrenheit thermometer plummet to 0° (−17.8°C).[113] The sudden thaw of early February might have started to dissolve the ice in Cowper's ink but it brought new forms of misery across the Channel. In Paris heavy snow and rapid thawing had already drowned women and children in the first days of 1784, when the book publisher Nicolas Ruault claimed that the capital had been turned into a new 'Venice without gondolas or gondoliers'. Yet by 25 February 1784 Ruault wrote to his brother in Normandy to let him know that the snows were back in such quantity that huge mounds of ice blocked the streets like 'hay stacks on a prairie' and, less picturesquely, that firewood supplies were now running dangerously low.[114]

111. William Cowper to Reverend William Unwin, Sunday 7 September 1783, *Electronic Enlightenment scholarly edition of correspondence*, http://dx.doi.org.sheffield.idm.oclc.org/10.13051/ee:doc/cowpwiOU0020156a1c (last accessed 18 December 2018); William Cowper to John Newton, Monday 8 September 1783, *Electronic Enlightenment scholarly edition of correspondence*, http://dx.doi.org.sheffield.idm.oclc.org/10.13051/ee:doc/cowpwiOU0020159a1c (last accessed 18 December 2018).

112. William Cowper to John Newton, Tuesday 13 January 1784, *Electronic Enlightenment scholarly edition of correspondence*, http://dx.doi.org.sheffield.idm.oclc.org/10.13051/ee:doc/cowpwiOU0020199a1c (last accessed 18 December 2018).

113. See Scarth, *Vulcan's fury*, p.119; White, *The Natural history and antiquities of Selbourne*, p.298, 293.

114. Nicolas Ruault, *Gazette d'un Parisien sous la Révolution: lettres à son frère, 1783-1796* (Paris, 1976), p.36-37.

Eighteenth-century explanations for volcanogenic weather

All of these personal accounts offer much more than a touch of local colour to the climate devastation wrought by Laki. Taken collectively, they may even represent the first glimmers of a modern ecological consciousness lit by the destructive fires of volcanoes, even though volcanism itself was imperfectly understood. Hence the perturbing volcanic – or rather volcanogenic – phenomena of 1783-1784 gave rise to a variety of traditional and novel explanations across all orders of society. One particularly persistent popular belief was in the baleful influence of passing comets causing large-scale atmospheric disturbances. Already in 1773 in France a paper presented to the Académie des sciences by the astronomer Joseph Jérôme Lalande had sparked widespread panic when he speculated that a comet would soon collide with the Earth. Voltaire, writing to none other than Sir William Hamilton in the June of that year, mocked the terror gripping women of all social orders in Paris who had hurried to confession on hearing of Lalande's dire predictions – confession being, as the notorious wit put it, 'an infallible secret for diverting comets from their course'.[115] Newspapers in both Britain and France mention a renewed sense of alarm at a possible comet impact in the summer of 1783, given the sickly sulphurous veil drawn across the sky and lingering there to pernicious effect. Louis Cotte sought to dismiss this popular superstition. Writing in the *Journal des sçavans* he stated that no comet had been spotted by astronomers in the spring or summer of 1783, and therefore hurtling cosmic debris could not be the cause of these persistent dry fogs.[116] Nonetheless, conjectures on the power of comets to periodically warp the climate lasted well into the nineteenth century.[117]

As Cotte's reflections suggest, a number of French scientists had their telescopes keenly trained on the polluted skies in the summer of 1783. According to Emmanuel Garnier, members of the Paris Observatory flew kites into the louring atmosphere, hung with chunks of meat, to assess the vitiating effects of the haze on foodstuffs.

115. See Joseph Jérôme Lalande, *Réflexions sur les comètes qui peuvent approcher de la Terre* (Paris, Gibert, 1773); Voltaire, *Correspondence and related documents*, vol.124, p.27, Voltaire to Sir William Hamilton, 17 June 1773 (D18429).

116. Cotte, 'Extrait des observations météorologiques faites à Laon', p.48-49.

117. See, for example, Edouard Roche, *Recherches sur les offuscations du soleil et les météores cosmiques* (Paris, 1868), p.62-63.

The meat was brought to earth 'entirely corrupted'.[118] Kites had been effective instruments of meteorological investigation ever since Benjamin Franklin had used them in the 1750s to understand the electrical nature of lightning. As fate would have it, Franklin was in Paris in 1783 as a member of the United States delegation to the peace talks officially ending the American War of Independence. The following year in Passy on the outskirts of Paris, he committed to paper his own thoughts on the bizarre weather events of 1783-1784. His 'Meteorological imaginations and conjectures' were published in the proceedings of the Manchester Literary and Philosophical Society in late 1784 through the good offices of his correspondent Dr Percival. His brief article starts with a reflection on the extreme cold of the preceding winter, attributing it to the screening effects of the fogs which reduced the surface temperature of the Earth, leading to early frosts and persistent snow cover. He then advances two rather ambivalent propositions by way of conclusion. Either the choking haze itself was caused by intense meteor showers 'which are sometimes seen to kindle and be destroyed in passing our atmosphere, and whose smoke might be attracted and retained by our earth'; or its origins lay in the volcanic eruptions of Iceland. He cites in particular 'the vast quantity of smoke, long continuing to issue during the summer from *Hecla* in Iceland, and the other volcano which arose out of the sea near that island, which smoke might be spread by various winds over the northern part of the world'.[119]

Franklin was a polymath and Enlightenment hero, hence this last conjecture is often quoted as further evidence of his unique genius. He is frequently presented as being the first to fully grasp the large-scale climate impact of Laki and, by extension, that of other cataclysmic eruptions. Yet, as the above citation makes clear, Franklin offers this solution as one of two equally possible causes for the dry fogs. He also mistakes Laki for Hekla, then dormant, and overestimates the importance of the offshore eruption of Reykjaneshryggur in early May 1783. More importantly, his short text was written in May 1784, thus it is far from being the first explicitly to link Icelandic volcanism and the subsequent freak meteorological conditions. Already on 7 August

118. Emmanuel Garnier, 'Les brouillards du Laki en 1783: volcanisme et crise sanitaire en Europe', *Bulletin de l'Académie nationale de médecine* 195:4-5 (2011), p.1043-55 (1046).

119. Benjamin Franklin, 'Meteorological imaginations and conjectures', in *Philosophical and miscellaneous papers* (London, C. Dilly, 1787), p.82-85.

1783 the little-known French naturalist Jacques Antoine Mourgue de Montredon had presented a paper to the Académie royale des sciences in Montpellier which made the connection between volcanic activity in Iceland and the extreme weather events of the summer. His findings were based on his daily observations of the bizarre 'non-aqueous vapours', as he called them, which initially battened onto southern France on 17 June 1783 and lingered heavily there until 22 July.[120] However, instead of seeing these noxious fogs as issuing solely from Icelandic volcanoes, Mourgue traces a deadly geological fracture across western Europe, from Iceland to Calabria, emitting sulphurous gases locally through Saxony, Bavaria, Hungary and Austria too. In other words, the volcanic fogs do not sweep down from Iceland borne on stratospheric winds but are exhaled from the Earth's crust along a continent-wide 'trail of subterranean fire'.[121] He claims that the poisonous geological fault line does not stop in Europe either, but extends into the Middle East and North Africa. It is only in his later observations of a second wave of stifling haze from 12 August to 4 September that the naturalist locates their source exclusively in the renewed activity of 'new volcanoes close to mount Hecla' spewing flame and smoke and flooding local lands with vast lava flows.[122] Further north, other students of the natural world would confirm Mourgue's association of the extreme weather events with Icelandic volcanism long before Franklin did. Thus the German natural historian Johann L. Christ makes this claim in his *Von der ausserordenlichen Witterung des Jahres 1783* published in Frankfurt and Leipzig in late 1783, as does the young Icelandic theology student Sæmundur Magnússon Hólm in a short text that appeared in Copenhagen in early 1784.[123]

120. Jacques Antoine Mourgue de Montredon, 'Recherches sur l'origine et sur la nature des vapeurs qui ont régné dans l'atmosphère pendant l'été de 1783', in *Histoire de l'Académie royale des sciences* (Paris, Imprimerie royale, 1784), p.754-73 (756).

121. Mourgue de Montredon, 'Recherches sur l'origine et sur la nature des vapeurs', p.760-61.

122. Mourgue de Montredon, 'Recherches sur l'origine et sur la nature des vapeurs', p.761-62.

123. Johann L. Christ, *Von der ausserordenlichen Witterung des Jahres 1783* (Frankfurt and Leipzig, Heimannischen Buchhandlung, 1783); Sæmundur Magnússon Hólm, *Om jordbranden paa Island i aaret 1783* (Copenhagen, Reder Horrebow, 1784); cited in Thordarson and Self, 'Atmospheric and environmental effects of the 1783-1784 Laki eruption', p.2. Bertholon, writing in 1787, claims that another amateur scientist, the Benedictine dom Robert Hickmann, had similarly linked the 'dry and sulphurous fog' to 'volcanic upheavals in Iceland' 'as early as July

Nonetheless, it seems clear from Mourgue's speculations that contemporary observers were fooled by the slow spiralling of the sulphuric clouds over the continent, and that what issued violently from north-eastern Europe appeared to stem almost simultaneously from south-western Europe, originating just as much in the seismic upheavals of Calabria and Sicily as it did in the volcanic eruptions in Iceland. This view was heavily influenced by the terrible news that had emerged from Calabria and Messina in February 1783 when a catastrophic earthquake killed between 30,000 and 50,000 inhabitants of the region and left the landscape buckled and shattered.[124] Some connection to the bizarre meteorological events of the summer seemed highly probable to many contemporary weather-watchers. In addition to Mourgue's speculations, Gilbert White also linked the devastating tremors in southern Italy to the later freak weather of the summer. This thesis was given further credibility by the Italian correspondent of the widely read *Journal des sçavans* Michele Torcia, who maintained that the atmospheric veil was thickest and most sulphurous over the epicentre of the Calabrian quake. In turn, both Jean-Louis Giraud-Soulavie and Louis Cotte propagated this opinion in France.[125] In fact, Cotte elaborated on it, developing a holistic reading of the volcanogenic weather that integrated meteorological, seismic and electrical theories in its reasoning. He writes in his 'Observations météorologiques' for July 1783:

> The sun began by pumping out all the humid vapours that had saturated the earth in the heavy winter rains; hence wet and cold fogs; the earth, now dried out by this strong and rapid evaporation, cracked in many places, releasing a quantity of sulphurous and other exhalations, of electrical matter too, stirred up by the terrible earthquake in Sicily and Calabria; hence these dry fogs and intense heat.[126]

1783', publishing his claim in the *Journal encyclopédique* of the same year. See Bertholon, *De l'électricité des météores*, vol.2, p.134.

124. See Quenet, *Les Tremblements de terre*, p.472.

125. White, *The Natural history and antiquities of Selbourne*, p.302; Torcia, 'Lettre sur le tremblement de terre de Calabre', p.840. Cotte, 'Extrait des observations météorologiques faites à Laon', p.48; and Jean-Louis Giraud-Soulavie, 'Lettre de M. l'abbé Giraud-Soulavie au R. P. Cotte, de l'Oratoire, curé de Montmorency: observations physiques sur un nuage apparent observé en Bourgogne', *Journal de Paris* 202-203 (21-22 July 1783). Cited in Grattan *et al.*, 'Volcanic air pollution and mortality in France 1783-1784', p.651.

126. Cotte, 'Extrait des observations météorologiques faites à Laon', p.48.

As in later popular 'electrical' conceptions of geological and meteorological phenomena, the one element that gets lost in Cotte's explanation of the dry fogs of 1783 is Icelandic volcanoes.

More time was perhaps needed for eighteenth-century empirical scientists to understand Laki and its expansive, deadly fallout. So we will leave the subject with an image of the Icelandic physician and naturalist Sveinn Pálsson walking the gnarled, solidified fissure of Laki in the summer of 1794. At the very time when Vesuvius was erupting violently and the revolutionary Terror was at its dreadful height in France, Pálsson took both a closer and a longer historical view of volcanic havoc. Eleven years after the calamitous explosion of the Laki craters, he found 'a column of smoke still arising from certain parts of the lava; and some of the rents were filled with hot water'.[127] The principal tract of molten rock was just over a mile wide, now hardened into a smooth channel. Heading towards the country's largest volcano Öræfajökull – where he would make the first recorded ascent of its highest peak on the crater rim at Hvannadalshnúkur – Pálsson was particularly attentive to the slow but constant shifts in the stark Icelandic topographies. On 11 August 1794, as he looked down on the Kvísker glacier from the summit of volcanic Öræfajökull, he saw more timeless geological processes at work in and around Laki. It was a dispassionate interrogation of the desolate landscape which confirmed him in his remarkable intuition about the way in which glaciers creep over time, analogous in their stealthy, smothering advance to tar or pitch.[128] Thus, beyond the urgent conjectures fuelled by the cataclysmic upheavals of Laki in 1783, volcanic Iceland continued to offer a uniquely inspiring, dynamic terrain for its earth scientists, explorers and tourists, as it would into the nineteenth century and as it still does today.

127. Cited in Ebenezer Henderson, *Iceland, or the Journal of a residence in that island, during the years 1814 and 1815*, 2 vols (Edinburgh, 1818), vol.1, p.286. Pálsson had left the manuscript of his 1792-1794 volcanic journey across Iceland with Henderson when they met in 1815.

128. Sveinn Pálsson, *Draft of a physical, geographical, and historical description of Icelandic ice mountains on the basis of the journey to the most prominent of them in 1792-1794* (Reykjavik, 2004), p.60-68. See also Helgi Björnsson, *The Glaciers of Iceland: a historical, cultural and scientific overview* (Paris and Amsterdam, 2017), p.151-59.

Conclusion

The volcano constitutes the mountain as event. While Etna smoking serenely is also a sublime object of contemplation on a par with Rousseau's Alps, its smoke nonetheless signifies another order of awe. The volcano represents cataclysmic power, and thus modifies its relationship with its beholder. It is the mountain not only as imposing spectacle but also, troublingly, as unpredictable, overwhelming devastation. It is the solidity of black rock turned into liquid fire, the eternal and immovable suddenly set in terrifying motion. So the volcano is about the eruptive event, large or small. That event may have happened aeons ago and be experienced only as the charred traces of geophysical violence, as with Faujas de Saint-Fond scrambling through the scoria-filled craters of otherwise green Auvergne. More often in the eighteenth century, volcanic prehistory is brought dizzyingly into the present. It emerges in the uncovered ruins of Pompeii and Herculaneum. It is discerned in Buffon's notion of geological 'actualism'. Most of all, it erupts violently into contemporary European consciousness with each fiery blast of Vesuvius, Etna, Hekla or Laki. These are relayed successively and breathlessly around Europe in text and image or are experienced as sublime spectacle on the Grand Tour, in deadly havoc in Iceland or in baleful climate shifts farther afield.

The volcano is at once past and present. It stands as an eternal reminder of mankind's own ephemerality. Generations pass, the smoking peak endures. As such, it is a place of life and death, prompting its eighteenth-century observers to celebrate common appetites to the full and to contemplate solemnly humanity's shared mortality. Even as it shapes neo-classical sensibilities among tourists pondering the ruins of Pompeii, Vesuvius becomes a site of lively enlightened sociability, its slopes streaming with scientists, artists, merchants, tourists and guides of many nationalities and different languages. This is exemplified in Roland de La Platière's seamless weaving together of classical reflections and reckless picnicking on the volcano's smoking lava fields. The eighteenth-century visitor revels in the perturbing, yet exhilarating, antitheses of the volcano: it is contemplation and action; an explosive peak that levels socially; one which humbles in its lesson of our common fate, yet at the same time incites each tourist to measure his or her vital forces against it. Little wonder then that, as it intensifies a feeling of individualism in many of its observers, the volcano itself is personified as a fire-spewing giant. In this guise, it inspired ancient cults of priapic sexual intemperance just as it appeared to push more modern local appetites to bulimic excess, exemplified by the rite of the Neapolitan 'cuccagna'. What makes the volcano, specifically Vesuvius, so powerful an experience in the eighteenth century is not just its intense concentration of these contradictory thoughts and feelings. It is also the intimation that sombre reflections on our common mortality and riotous calls to indulge every appetite may be reconciled *there as nowhere else.*

The same keen curiosity that drove eighteenth-century tourists to Vesuvius and Etna had for a long time fuelled scientific interest in volcanism. In the eighteenth century this curiosity informed the writings of the two modern 'Plinies' who most influenced scientific thinking about the workings of volcanoes. That is, the holistic theorising of Buffon and the intrepid empiricism of Sir William Hamilton. Early modern volcanology favoured analogies drawn from other disciplines in order to understand the terrible magnitude of volcanic power and the unfathomable geological depths from which it burst. Hence the volcano is a telluric vent or chimney (architecture); it is an explosive reaction of pyrite and seawater (chemistry); or it is the violent release of subterranean electricity (physics). Yet these largely abstract explanations for volcanic activity were increasingly superseded by findings in the field. Initially driven by an economic interest in mapping mineral and ore deposits, scientists roaming among the extinct volcanoes of central France started to identify the distinct basalt topographies of ancient volcanism. Among cosmopolitan scientific networks of the mid- to late eighteenth century, such regular black outcrops were quickly recognised across the continent, revising the importance of volcanoes as forces forming and defining the Earth's surface. Yet, from such shared empirical beginnings, the same scientists split into rival schools of thought, Neptunists versus Plutonists, extrapolating divergent theories from the new hauls of evidence gathered in the field. Thus, while it had become a privileged site for often perilous scientific fieldwork, the volcano remained stubbornly an emblem of primitive, elemental powers, chief proof of the Plutonists' claims for the primary earth-shaping forces of fire.

However, the volcano appealed not only to scientific reason but also to atavistic memory and exalted imagination. From Iceland to Santorini, devastating volcanic eruptions were burned into the ancestral memories of European peoples, however much enlightened scientists sought to find more reassuringly rational causes for them. Consciously or not, these same early volcanologists supplemented and enhanced the powers of reason with their own excited imagination as they sought to grasp the incommensurable forces at work in volcanic eruptions. And what was true for eighteenth-century scientists was truer still for contemporary aesthetic theorists and artists. In the work of Burke and Kant, the imagination is at once extolled and exceeded in confronting the volcano's terrific explosions and outpourings. The name of this thrilling imaginative alchemy, turning raw fear into refined delight, is the *sublime*. The volcano thus becomes a major testing ground for the intellectual framework provided by theorists of natural sublimity. Increasing numbers of artists flocked to Vesuvius, often in the entourage of wealthy patrons or in the company of vagabond writers, in order to pit their art against nature, to capture on canvas the inconceivable power of the volcano. In attempting to represent volcanic nature, eighteenth-century art often only succeeded in domesticating it, framing and curbing the excessive sublime in picturesque compositions. Once tamed, the volcano became a more malleable aesthetic figure, reproduced artificially in firework displays and in stately parks, as at Wörlitz in eastern Germany. While the volcano continued to disturb deeply its observers with its uncanny extrusions of incalculable amounts of molten matter, its sublimity was increasingly assimilated to that of huge industrial forges and glasshouses. In turn, tourists, in their 'eruption narratives', appropriated the volcanic sublime for themselves, transferring the unimaginable sublimity of the spectacle to the imaginative writer-spectator who dared to confront it.

For yet other Enlightenment thinkers, the volcano posed the problem of its place in God's creation. It had long been associated with visions of hell as it was now with modern industrial furnaces. But, as contemporary science presented the erupting volcano as more chemical than infernal, some theologians attempted to align volcanism more directly with biblical cosmogony and eschatology, to read nature literally as scripture. Thus physico-theological writers, such as Burnet, saw in the volcano both the fallen world of Genesis and the fiery resurrection of Revelation. Nonetheless, as the century wore on, the volcano proved more legible as a feature of natural theology than as a cipher of Christian doctrine; volcanism appeared increasingly as an immanent wonder of the natural world, not as proof of exceptional divine intervention in earthly matters. As an integral part of a beneficent deity's cosmic plan, the volcano was thus recuperated by providentialists and theists in order to demonstrate its constructive role in long-term natural renewal beyond its immediate violence and devastation. Deists, such as Hamilton and Voltaire, agreed that the volcano was ultimately 'good'; but they were more interested in its anti-clerical properties, as material proof of geological deep time which clearly refuted the biblical chronology. It also offered them a glaringly risible example of contemporary Catholic superstition in the clamorous and ineffectual parading of religious relics to ward off eruptions. As the bigoted priesthood led a benighted people through Naples in response to Vesuvius erupting, enlightened observers also saw a convergence of volcanic violence and popular unrest. For their part, they preferred to view the volcano as a source of light, rather than heat – as a dreadful fire to be mastered in Promethean style. Nonetheless, a countervailing vision of the volcano sought less to control its jetting fires than to unite with them, as legend claimed Empedocles had done. And, in the last decades of the eighteenth century, it was this latter vision which began to win out. The volcanic sublime, the volcanic figure of regenerative rupture, the volcanic power of the insurgent people, and the temptation of transcendent Empedoclean self-sacrifice in the volcano's flames – all these combined in an intoxicating mix to make the volcano nature's principal *revolutionary*.

The French Revolution confirmed spectacularly the politicisation of the volcano. Volcanic figures of speech became part of the aestheticisation of revolutionary politics, and whoever controlled the language of the revolutionary sublime appeared to shape the direction of the Revolution itself. There were two principal functions to volcanic metaphors in the Revolution: to warn of the dangerous latency of one's political opponents or to celebrate the irresistible might of one's own revolutionary faction. These represent two phases in the eruptive cycle: a terrible imminent eruption or a tremendous eruption in action. Actors and writers in the Revolution sought either to play Prometheus and bridle volcanic power or to join Empedocles in a headlong immersion in it. In the period of the revolutionary Terror (1793-1794), it was the latter Empedoclean politics which prevailed. The figure of the gushing, fiery volcano sweeping all before it was superimposed on the vehement, violent radicalism of the 'Montagne', the faction of extreme-left deputies sitting in the Convention and dominant in the Jacobin Club. Ideologically in this period, however, volcanic rhetoric proved pliant enough to accommodate both the materialist-atheist Terror of the dechristianisation campaign of late 1793 and the Rousseau-inspired deism of Robespierre in the spring of 1794. In June 1794, at the height of a continent-wide war between a coalition of European monarchies and republican, regicide France, Vesuvius itself began spouting fountains of fire and enormous mushroom clouds of black ash. It thus appeared to many observers in France and beyond that the volcanic conflagration of Revolution was set to engulf all of Europe. Yet these volcanic upheavals ended up imitating the more famous figure of the Revolution as Saturn 'eating his own children'. They ultimately consumed those who had thought to train the volcano's dreadful explosions on their enemies both at home and abroad.

The volcano in eighteenth-century Europe is then a significant point of convergence for geophysics and geopolitics. Nowhere was this more manifest than in Iceland, whose founding tenth-century parliament had convened in the volcano-riven tectonic trench at Þingvellir. By the mid-eighteenth century, however, Iceland had become a blasted volcanic outpost of the kingdom of Denmark. Its most famous volcano, Hekla, was conventionally cited in scientific studies of the time, but it was mainly local Scandinavian explorers who scoured the island to put seventeenth-century theories of volcanism to the test. What they discovered was the unique nature of Icelandic volcanic activity: fiery belts of volcanic vents overlain with glacial ice caps, an explosive collision of fire and ice. In the latter half of the eighteenth century this was confirmed by an increasing number of western European explorers of the island who often professed a disinterested enlightened quest for knowledge as cover for more interested, colonial ventures. Volcanic Iceland thus became important in the imperial politics of the late eighteenth century, connoted in Sir Joseph Banks' disingenuous claims in 1772 of being the first to 'conquer' Hekla. Yet it was the little-known Laki volcanic fissure, erupting violently in 1783, that was to prove how volcanic Iceland could impact significantly on the societies of western Europe. As the site of one of the most deadly eruptions in human history, Laki has since been extensively studied. However, the harrowing accounts of its local devastation tell a lot about contemporary Icelandic belief systems, the politics of subsistence farming, and the limits of physical resilience in bleak volcanic terrain. Farther afield, Laki cast a choking sulphurous veil over western Europe that spread as far south as North Africa. Its deleterious climate effects provoked both a recourse to superstitious readings of the weather and inductive meteorological observation. Laki demonstrated once again how volcanism represented an explosive threat to European societies, especially those cohabiting with active craters, and constituted an invaluable spur to revise traditional understandings of the world.

In eighteenth-century Europe, the volcano was thus a near-ubiquitous figure informing almost every field of human endeavour. As this study has attempted to show, erupting volcanoes were a physical reality for many Europeans of the time. For many others, they represented the very image of overwhelming natural power, whether this was ultimately attributed to spiritual or material causes. As such, the volcano proved an effective and versatile 'tool for thinking' in a century ushering in modernity on several fronts: continental tourism, new earth sciences, the sublime and picturesque in art, industrial and political revolution, and the conception of the modern nation-state. It also gives us, in the twenty-first century, a privileged site (as both topography and topos) at which we can reconnect disparate and divided fields of research across the sciences and the humanities. The volcano is clearly transnational, but this study has also sought to show how it is fundamentally transdisciplinary. In their often extreme material and symbolic transformations, the volcanic 'environments' explored in this book – mountains like Vesuvius, regions like Auvergne, ruins like Pompeii, cities like Naples or revolutionary Paris, the island of Iceland – do not only modify adjoining spaces of human activity but also reconfigure the knowledge formed at the interface of the human and the volcanic. This is one of the ways in which a study of volcanoes in eighteenth-century Europe might also be considered an essay in environmental humanities.

Bibliography

Works written before 1850

'An account of the eruption of Mount Vesuvius in 1767, in a letter from an English gentleman residing at Naples, to John Morgan, M. D. Prof. Med. Coll. Philad.', *Transactions of the American Philosophical Society* 1 (January 1769-January 1771), p.281-85.

Addison, Joseph, *Remarks on several parts of Italy, etc. in the years 1701, 1702, 1703* (London, Jacob Tonson, 1705).

Ætna, ed. Robinson Ellis (Oxford, 1901).

Amato, Gaetano d', *Divisamento critico sulle correnti opinioni intorno a fenomeni del Vesuvio, e degli altri vulcani* (Naples, Stamperia Abbaziana, 1756).

Anderson, Johann, *Nachrichten von Island, Grönland und der Strasse Davis, zum wahren Nutzen der Wissenschaften und der Handlung* (Hamburg, Georg Christian Grund, 1746).

Baculard d'Arnaud, François-Thomas-Marie de, *Euphémie, ou le Triomphe de la religion: drame*, 2nd edn (Paris, Le Jay, 1768).

Baldini, Filippo, *De' sorbetti: saggio medico-fisico* (Naples, n.n., 1775).

Banks, Sir Joseph, 'The Iceland journal of Sir Joseph Banks, Part I: 12 July-6 September 1772', in *Sir Joseph Banks, Iceland and the North Atlantic 1772-1820: journals, letters and documents*, ed. Anna Agnarsdóttir (London, 2016), p.45-92.

–, 'The Iceland journal of Sir Joseph Banks, Part II: 17 September-22 October 1772', in *Sir Joseph Banks, Iceland and the North Atlantic 1772-1820: journals, letters and documents*, ed. Anna Agnarsdóttir (London, 2016), p.93-114.

Barère, Bertrand, *Mémoires de B. Barère, membre de la Constituante, de la Convention, du Comité de salut public, et de la Chambre des représentants*, 4 vols (Paris, 1842-1844).

–, *Rapport du citoyen Barère, sur la reprise de Toulon par l'armée de la République* (Paris, Imprimerie nationale, 1793).

–, *Rapport fait au nom du Comité de salut public, sur l'état de la fabrication révolutionnaire du salpêtre et de la poudre* (Paris, Imprimerie nationale, 1794).

–, *Rapport sur les crimes de l'Angleterre envers le peuple français, et sur ses attentats contre la liberté des nations* (Paris, Imprimerie nationale, 1794).

Beckford, William, *Biographical memoirs of extraordinary painters* (London, J. Robson, 1780).

Bergeret de Grancourt, Pierre-Jacques-Onesyme, *Bergeret et Fragonard: journal inédit d'un voyage en Italie 1773-1774* (Paris, 1895).

Bernardin de Saint-Pierre, Jacques-Henri, *Les Etudes de la nature*, 3 vols (Paris, Imprimerie de Monsieur, 1784).

–, *Harmonies de la nature*, ed. Louis Aimé-Martin, 3 vols (Paris, 1815).

–, *Paul et Virginie* (Paris, 1989).

Bertholon, Pierre-Nicolas, *De l'électricité des météores*, 2 vols (Paris, Croullebois, 1787).

–, 'Mémoires sur un para-tremblement de terre et un para-volcan', *Journal encyclopédique* 3 (1779), p.123-32, 323-31.

Bertuch, Friedrich Justin, *Bilderbuch für Kinder*, 12 vols (Weimar, 1795-1824).

Billaud-Varenne, Jacques-Nicolas, *Mémoires de Billaud-Varennes, ex-conventionnel, écrits au Port-au-Prince en 1818*, 2 vols (Paris, 1821).

Blake, William, *Poetical sketches* (London, n.n., 1783).

Boileau-Despréaux, Nicolas, *Traité du sublime*, in *Œuvres complètes* (Paris, 1966), p.331-440.

Borelli, Giovanni Alfonso, *Historia ed meteorologia incendii Aetnaei anni 1669* (Naples, Egidio Longo, 1670).

Boullée, Etienne-Louis, online exhibition at the Bibliothèque nationale de France, http://expositions.bnf.fr/boullee/arret/d4/index.htm (last accessed 10 December 2018).

Bourguet, Louis, *Traité de pétrifications* (Paris, Briasson, 1742).

Bourrit, Marc-Théodore, *Nouvelle description générale et particulière des glacières, vallées de glace et glaciers qui forment la grande chaîne des Alpes de Suisse, d'Italie & de Savoye*, 3 vols (Geneva, Paul Barde, 1785).

Braccini, Giulio Cesare, *Dell'incendio fattosi nel Vesuvio a XVI. di Dicembre M.DC.XXXI. e delle sue cause, ed effetti* (Naples, Secondino Roncagliolo, 1632).

Breislak, Scipione, *Voyages physiques et lythologiques dans la Campanie*, 2 vols (Paris, 1801).

Bromley, William, *Several years travels through Portugal, Spain, Italy, Germany, Prussia, Sweden, Denmark and the United Provinces* (London, n.n., 1702).

Brydone, Patrick, *A Tour through Sicily and Malta, in a series of letters to William Beckford*, 2 vols (London, W. Stahan; T. Cadell, 1773).

Buache, Philippe, 'Essai de géographie physique, où l'on propose des vûes générales sur l'espèce de charpente du globe, composée des chaînes de montagnes qui traversent les mers comme les terres, avec quelques considérations particulières sur les différens bassins de la mer, & sur sa configuration intérieure', *Mémoires de l'Académie royale des sciences* (1752), p.399-416.

Buffon, Georges-Louis Leclerc, comte de, *Histoire naturelle, générale et particulière*, 36 vols (Paris, Imprimerie royale, 1749-1788).

–, *Histoire naturelle: supplément*, vol.1 (Paris, Imprimerie royale, 1774).

–, *Histoire naturelle: supplément*, vol.2 (Paris, Imprimerie royale, 1775).

–, *Histoire naturelle: supplément*, vol.5: *Des époques de la nature* (Paris, Imprimerie royale, 1778).

Burke, Edmund, *A Philosophical enquiry into the origin of our ideas of the sublime and the beautiful* (London, 1958).

–, *Reflections on the Revolution in France* (Oxford, 1999).

Burnet, Thomas, *The Sacred theory of the Earth*, 3rd edn (London, Walter Kettilby, 1697).

Calonne, Charles Alexandre de, *De l'état de la France tel qu'il peut et qu'il doit être* (London, n.n., 1790).

Cartaud de La Villate, François, *Essai historique et philosophique sur le goust* (Amsterdam, n.n., 1736).

Casanova de Seingalt, Jacques, *Histoire de ma vie*, 12 vols (Wiesbaden, 1960-1962).

Chamfort, Sébastien-Roch Nicolas, *Œuvres complètes de Chamfort*, 5 vols (Paris, 1824-1825).

Chénier, André, *Œuvres complètes* (Paris, 1950).

Christ, Johann L., *Von der ausserordenlichen Witterung des Jahres 1783* (Frankfurt and Leipzig, Heimannischen Buchhandlung, 1783).

Cizos-Duplessis, François, *Les Peuples et les rois, ou le Tribunal de la raison* (Paris, Barba, 1794).

Clairvaux, Bernard de, *De miraculis* (1180).

Corrado, Vincenzo, *Il credenziere di buon gusto* (Naples, n.n., 1778).

–, *Il cuoco galante* (Naples, Stamperia Raimondiana, 1773).

Cotte, Louis, 'Extrait des observations météorologiques faites à Laon, par ordre du roi, pendant le mois de juillet 1783', *Le Journal des sçavans* (1784), p.48-52.

–, 'Mémoire: sur les brouillards extraordinaires des mois de juin & de juillet 1783', *Journal de physique, de chimie, d'histoire naturelle et des arts* 23 (September 1783), p.202-205.

Cowper, William, *Poems by William Cowper* (London, J. Johnson, 1782).

Daquin, Joseph, *Essai météorologique sur la veritable influence des astres, des saisons et des temps* (Chambéry, Gorrin, 1784).

de Gouges, Olympe, *Ecrits politiques*, 2 vols (Paris, 1993).

Delisle de Sales, Jean-Baptiste-Claude, *De la philosophie de la nature*, 3 vols (Amsterdam, Arkstée & Merkus, 1770).

Della Torre, Giovanni Maria, *Histoire et phénomènes du Vésuve* (Paris, Hérissant, 1760).

–, *Storia e fenomeni del Vesuvio* (Naples, Giuseppe Raimondi, 1755).

de Luc, Jean-André, *Lettres physiques et morales, sur les montagnes et sur l'histoire de la Terre et de l'homme* (The Hague, Detune, 1778).

de Maillet, Benoît, *Telliamed, ou Entretiens d'un philosophe indien avec un missionnaire françois sur la diminution de la mer, la formation de la terre, l'origine de l'homme, &c.*, 2 vols (Amsterdam, L'Honoré & Fils, 1748).

Denon, Dominique Vivant, *Voyage en Sicile* (Paris, Didot l'Aîné, 1788).

Derham, William, *Physico-theology, or a Demonstration of the being and attributes of God, from his works of creation* (London, W. Innys, 1713).

Desbarreaux, Citoyen, *Les Potentats foudroyés par la Montagne et la Raison, ou la Déportation des rois de l'Europe* (Toulouse, P. Francés, 1793).

Desmarest, Nicolas, 'Mémoire sur l'origine & la nature du basalte à grande colonnes polygones, déterminées par l'histoire naturelle de cette pierre, observée en Auvergne', *Mémoires de l'Académie royale des sciences* (1774), p.705-75.

Dibdin, Thomas, *Songs, recitatives, &c. in the volcano, or the Rival harlequins* (London, J. Barker, 1799).

Diderot, Denis, *Salon de 1767*, in *Œuvres complètes*, ed. J. Assézat and M. Tourneux, 20 vols (Paris, 1875-1877), vol.11, p.3-367.

–, and Jean D'Alembert (ed.), *Encyclopédie, ou Dictionnaire raisonné des sciences, des arts et des métiers*, 35 vols (Paris, Briasson, 1751-1780).

Diogenes Laertius, *Lives of eminent philosophers*, 2 vols (London, 1925).

Dolomieu, Déodat de, *Mémoire sur les îles ponces, et Catalogue raisonné des produits de l'Etna, pour servir à l'histoire des volcans, suivis de la description de l'éruption de l'Etna, du mois de juillet 1787* (Paris, Cuchet, 1788).

–, *Mémoire sur les tremblemens de terre de la Calabre pendant l'année 1783* (Rome, Antoine Fulgoni, 1784).

–, 'Rapport fait à l'Institut national, par le C. Dolomieu, ingénieur des mines, sur ses voyages de l'An v et de l'An vi', *Journal des mines* 46-47 (1798), p.385-402, 405-32.

Dortous de Mairan, Jean-Jacques, 'Nouvelles recherches sur la cause générale du chaud en été & du froid en hiver, en tant qu'elle se lie à la chaleur interne & permanente de la Terre', *Histoire de l'Académie royale des sciences, année M. DCCLXV* (Paris, Imprimerie royale, 1768), p.143-267.

Duclos, Charles Pinot, *Voyage en Italie, ou Considérations sur l'Italie* (Paris, Buisson, 1791).

Dupaty, Charles-Marguerite Mercier, *Lettres sur l'Italie, en 1785*, 2 vols (Paris, De Senne, 1788).

Eidophusikon, or Various imitations of natural phenomena, represented by moving pictures, invented and painted by Mr. de Loutherbourg (London, n.n., 1782).

Fabris, Pietro, *Raccolta di varii vestimenti ed arti del regno di Napoli* (Naples, n.n., 1773).

Faujas de Saint-Fond, Barthélemy, *Minéralogie des volcans, ou Description de toutes les substances produites ou rejetées par des feux souterrains* (Paris, Cuchet, 1784).

–, *Recherches sur les volcans éteints du Vivarais et du Velay* (Grenoble, Joseph Cuchet, 1778).

–, *Voyage en Angleterre, en Ecosse et aux îles Hébrides*, 2 vols (Paris, H. J. Jansen, 1797).

Fénelon, François de Salignac de La Mothe, *Demonstration de l'existence de Dieu, tirée de la connoissance de la nature, & proportionnée à la foible intelligence des plus simples*, 2nd edn (Paris, Jacques Estienne, 1713).

Féraud, Jean-François, *Dictionnaire critique de la langue française*, 3 vols (Marseille, Mossy, 1787-1788).

Ferber, Johann Jakob, *Briefe aus Wälschland über natürliche Merkwürdigkeiten* (Prague, Wolfgang Gerle, 1773).

Ferrara, Francesco, *Descrizione dell'Etna con la storia delle eruzioni e il catalogo dei prodotti* (Palermo, 1818).

Finch, Heneage, earl of Winchilsea, *A True and exact relation of the late prodigious earthquake & eruption of Mount Aetna, or Monte-Gibello* (London, T. Newcombe, 1669).

Fortis, Alberto, *Saggio d'osservazioni sopra l'isola di Cherso ed Osero* (Venice, Gaspare Storti, 1771).

–, *Viaggio in Dalmazia*, 2 vols (Venice, Alvise Milocco, 1774).

Franklin, Benjamin, 'Meteorological imaginations and conjectures', in *Philosophical and miscellaneous papers* (London, C. Dilly, 1787), p.82-85.

Galanti, Giuseppe Maria, *Breve descrizione della città di Napoli e del suo contorno* (Naples, presso li Soci del Gabinetto Letterario, 1792).

Galiani, Ferdinando, *Spaventosissima descrizione dello spaventoso spavento che ci spaventò tutti coll'eruzione del Vesuvio la sera degli otto d'agosto 1779, ma (per grazia di Dio) durò poco* (Naples, 1825).

Gilpin, William, *Observations on the River Wye, and several parts of South Wales, etc., relative chiefly to picturesque beauty, made in the summer of 1770* (London, R. Blamire, 1782).

Gimma, Giacinto, *Della storia naturale della gemme, delle pietre, e di tutti i minerali, ovvero Della fisica sotterranea*, 2 vols (Naples, Gennaro Muzio, 1730).

Giraud-Soulavie, Jean-Louis, *Histoire naturelle de la France méridionale*, 7 vols (Paris, Jean-François Quillot *et al.*, 1780-1784).

–, 'Lettre de M. l'abbé Giraud-Soulavie au R. P. Cotte, de l'Oratoire, curé de Montmorency: observations physiques sur un nuage apparent observé en Bourgogne', *Journal de Paris* 202-203 (21-22 July 1783).

Goethe, Johann Wolfgang von, *Italian journey*, translated by W. H. Auden and Elizabeth Mayer (London, 1962).

–, *Italienische Reise* (Wiesbaden, 1959).

Goree, Father, 'A relation of a new island which was raised up from the bottom of the sea on the 23rd of May 1707', *Philosophical transactions* 27 (1711), p.354-75.

Guettard, Jean-Etienne, 'Mémoire sur quelques montagnes de la France qui ont été des volcans', *Mémoires de l'Académie royale des sciences* (1752), p.27-59.

Gulliver revived: containing singular travels, campaigns, voyages and adventures in Russia, the Caspian Sea, Iceland, Turkey, Egypt, Gibraltar, up the Mediterranean, on the Atlantic Ocean, and through the centre of Mount Etna into the South Sea [...] *by Baron Munchausen* (Dublin, P. Byrne, 1788).

Hamilton, Sir William, 'An account of the late eruption of Mount Vesuvius [June 1794]', *Philosophical transactions* 85 (1795), p.73-116.

–, *Campi Phlegraei: observations on the volcanos of the Two Sicilies*, 2 vols (Naples, n.n., 1776).

–, *Observations on Mount Vesuvius, Mount Etna, and other volcanos: in a series of letters addressed to the Royal Society* (London, T. Cadell, 1772).

–, 'The present state of Mount Vesuvius, with an account of a journey into the province of Abruzzo, and a voyage to the island of Ponza', *Philosophical transactions* 76 (1786), p.365-81.

–, *Supplement to the 'Campi Phlegraei', being an account of the great eruption of Mount Vesuvius in the month of August 1779* (Naples, n.n., 1779).

Hartig, Franz von, *Lettres sur la France, l'Angleterre, et l'Italie* (Geneva, n.n., 1785).

Henderson, Ebenezer, *Iceland, or the Journal of a residence in that island, during the years 1814 and 1815*, 2 vols (Edinburgh, 1818).

Hereford, Charles John Ann, *The History of France, from the first establishment of that monarchy, brought down to, and including a complete narrative of the late Revolution*, 3 vols (London, C. & C. Kearsley, 1795).

Herschel, William, 'An account of three volcanos in the moon', *Philosophical transactions* 77 (1787), p.229-32.

Holbach, Paul-Henri Thiry, baron d', *Système de la nature* (London, n.n., 1770).

Hölderlin, Friedrich, *The Death of Empedocles: a mourning-play*, ed. and translated by David Farrell Krell (New York, 2008).

–, *Sämtliche Werke*, ed. Friedrich Beissner, 15 vols (Stuttgart, 1943-1985).

Holm, Sæmundur Magnússon, *Om jordbranden paa Island i aaret 1783* (Copenhagen, Reder Horrebow, 1784).

Hooke, Robert, *The Posthumous works of Robert Hooke* (London, Sam. Smith and Benj. Walford, 1705).

Hooker, Sir William Jackson, *Journal of a tour in Iceland, in the summer of 1809*, 2nd edn, 2 vols (London, 1813).

Horace, *Satires, epistles and ars poetica* (London, 1970).

–, *Satires, epistles, the art of poetry* (London, 1926).

Horrebow, Niels, *The Natural history of Iceland: containing a particular and accurate account of the different soils, burning mountains, minerals, vegetables, metals, stones, beasts, birds and fishes, together with the disposition, customs, and manner of living of the inhabitants* (London, A. Linde, 1758).

Houël, Jean-Pierre, *Voyage pittoresque des isles de Sicile, de Malte et de Lipari*, 4 vols (Paris, Imprimerie de Monsieur, 1782-1787).

Hume, David, 'Of the rise and progress of the arts and sciences', in *Essays moral and political*, 2 vols (Edinburgh, A. Kincaid, 1742), vol.2, p.53-100.

Hutton, James, *Theory of the Earth, with proofs and illustrations*, 2 vols (Edinburgh, William Creech; London, Messrs Cadell, Junior, and Davies, 1795).

Kant, Immanuel, *Beobachtungen über das Gefühl des Schönen und Erhabenen*, in *Werkausgabe*, 12 vols (Frankfurt am Main, 1977), vol.2, p.821-84.

–, *Kritik der Urteilskraft*, in *Werkausgabe*, 12 vols (Frankfurt am Main, 1977), vol.10, p.69-456.

Kerguelen-Trémarec, Yves-Joseph de, *Relation d'un voyage dans la mer du Nord, aux côtes d'Islande, du Groenland, de Ferro, de Schettland, des Orcades & de Norwége, fait en 1767 & 1768* (Paris, Imprimerie de Prault, 1771).

Kinsky, Franz Joseph, 'Nachricht von einigen Erdbränden im Ellenbogner Kreise in Böhmen', *Abhandlungen einer Privatgesellschaft in Böhmen* 2 (1776), p.58-73.

Kircher, Athanasius, *Mundus subterraneus* (Amsterdam, Joannem Janssonium & Elizeum Weyerstraten, 1665).

Kirwan, Richard, *Geological essays* (London, T. Bensley, 1799).

Kotzebue, August von, *Erinnerungen von einer Reise aus Liefland nach Rom und Neapel*, 3 vols (Berlin, 1805).

La Dixmerie, Nicolas Bricaire de, *Les Deux âges du goût et du génie français, sous Louis XIV & sous Louis XV* (Amsterdam, Barthelemi Vlam, 1770).

Lalande, Joseph Jérôme, *Abrégé d'astronomie*, 2nd edn (Paris, Firmin Didot, 1795).

–, *Réflexions sur les comètes qui peuvent approcher de la Terre* (Paris, Gibert, 1773).

–, *Voyage d'un François en Italie, fait dans les années 1765 et 1766*, 8 vols (Paris, Desaint, 1769).

La Roque, Jacques-Joseph de, *Voyage d'un amateur des arts en Flandre, dans les Pays-Bas, en Hollande, en France, en Savoie, en Italie, en Suisse, fait dans les années 1775-76-77-78*, 4 vols (Amsterdam, n.n., 1783).

Ledoux, Claude-Nicolas, 'La ville de Chaux, cité idéale de Claude-Nicolas Ledoux', *Passerelle(s)*, http://passerelles.bnf.fr/albums/chaux/index.htm (last accessed 10 December 2018).

Legrand d'Aussy, Pierre Jean-Baptiste, *Voyage fait en 1787 et 1788, dans la ci-devant Haute et Basse Auvergne*, 3 vols (Paris, Imprimerie des sciences et arts, 1794).

Lémery, Nicolas, *Cours de chymie*, 10th edn (Paris, Jean-Baptiste Delespine, 1713).

Lesur, Charles-Louis, *La Veuve du républicain, ou le Calomniateur* (Paris, Crapelet, 1793).

Linguet, Simon-Nicolas Henri, *Histoire impartiale des jésuites, depuis leur établissement jusqu'à leur première expulsion*, 2 vols (n.p., n.n., 1768).

–, 'Italie: éruption du Vésuve du 8 aout 1779', in *Annales politiques, civiles et littéraires du dix-huitième siècle*, 19 vols (London, Brussels and Paris, n.n., 1777-1792), vol.7, p.26-42.

Lister, Martin, 'Three papers of Dr Martin Lyster, the first of the nature of earth-quakes, more particularly of the origine of the matter of them, from the pyrites alone'; 'The second paper of the same person concerning the spontaneous firing of the pyrites'; 'The third paper of the same person, concerning thunder and lightning being from the pyrites', *Philosophical transactions* 14 (1684), p.512-19.

Longinus, *Dionysius Longinus on the sublime: translated from the Greek, with notes and observations, and some account of the life, writings and character of the author, by William Smith* (London, J. Watts, 1739).

–, *On sublimity*, translated by Donald A. Russell (Oxford, 1965).

Lucretius, *De rerum natura* (London, 1955).

Macquer, Pierre Joseph, *Dizionario di chimica*, 10 vols (Naples, G.-M. Porcelli, 1784-1786).

Madival, J., E. Laurent *et al.* (ed.), *Archives parlementaires de 1789-1860: recueil complet des débats législatifs et politiques des chambres françaises*, 82 vols (Paris, 1862-1914).

Maistre, Xavier de, *Voyage autour de ma chambre* (Paris, 2003).

Maréchal, Sylvain, *Le Jugement dernier des rois*, in *Théâtre du XVIII*[e] *siècle*, ed. Jacques Truchet, 2 vols (Paris, 1972-1974), vol.2, p.1307-25.

–, *Voyages de Pythagore en Egypte, dans la Chaldée, dans l'Inde, en Crète, à Sparte, en Sicile, à Rome, à Carthage, à Marseille et dans les Gaules, suivis de ses lois politiques et morales*, 6 vols (Paris, Deterville, 1798-1799).

Martínez Moles, Francisco, *Dissertacíon physica: origen, y formacíon del terremoto, padecido el día primero di noviembre de 1755* (Madrid, Juan de San Martín, 1755).

Mayeur de Saint-Paul, François-Marie, *Le Chroniqueur désœuvré, ou l'Espion du boulevard du Temple*, 2nd edn (Londres [Paris?], n.n., 1782).

Mercier, Louis-Sébastier, *Tableau de Paris*, 2 vols (Paris, 1994).

Michelet, Jules, *Histoire de la Révolution française*, 2 vols (Paris, 1998).

Miller, Lady Anna Riggs, *Letters from Italy, describing the manners, customs, antiquities, paintings, etc. of that country in the years MDCCLXX and MDCCLXXI to a friend residing in France*, 3 vols (Dublin, Watson *et al.*, 1776).

Montesquieu, Charles-Louis de Secondat de, *Considérations sur les causes de la grandeur des Romains et de leur décadence* (Oxford, 2000).

Montlosier, François-Dominique de Reynaud de, *Essai sur la théorie des volcans d'Auvergne*, 2nd edn (Paris and Clermont-Ferrand, 1802).

Morgan, Thomas, *Physico-theology, or a Philosophical-moral disquisition concerning human nature, free agency, moral government, and divine providence* (London, T. Cox, 1741).

Moro, Anton Lazzaro, *De' crostacei e degli altri marini corpi che si truovano su' monti* (Venice, Stefano Monti, 1740).

Mourgue de Montredon, Jacques Antoine, 'Recherches sur l'origine et sur la nature des vapeurs qui ont régné dans l'atmosphère pendant l'été de 1783', in *Histoire de l'Académie royale des sciences* (Paris, Imprimerie royale, 1784), p.754-73.

Ólafsson, Eggert, and Bjarni Pálsson, *Voyage en Islande*, translated by Gauthier de La Peyronie, 5 vols (Paris, 1802).

Oracle and public advertiser 18745 (Monday 14 July 1794).

Pálsson, Sveinn, *Draft of a physical, geographical, and historical description of Icelandic ice mountains on the basis of the journey to the most prominent of them in 1792-1794* (Reykjavik, 2004).

Paris, John Ayrton, *The Life of Sir Humphry Davy* (London, 1830).

Pennant, Thomas, *A Tour in Scotland, and voyage to the Hebrides 1772* (Chester, John Monk, 1774).

Peucer, Kaspar, *De dimensione terræ* (1550).

Piozzi, Hester Lynch Thrale, *Observations and reflections made in the course of a journey through France, Italy and Germany* (Dublin, H. Chamberlaine *et al.*, 1789).

Piron, Alexis, *La Métromanie, ou le Poète*, in *Théâtre du XVIIIe siècle*, ed. Jacques Truchet, 2 vols (Paris, 1972-1974), vol.1, p.1015-131.

Pixerécourt, René-Charles Guilbert de, *Belvéder, ou la Vallée de l'Etna* (Paris, 1818).

–, *La Tête de mort, ou les Ruines de Pompeïa* (Paris, 1828).

Pliny the Younger, *Letters and Panegyricus*, 2 vols (London, 1959).

Pluche, Noël-Antoine, *Le Spectacle de la nature, ou Entretiens sur les particularités de l'histoire naturelle*, 9 vols (Paris, veuve Estienne, 1732-1742).

Price, Uvedale, *An Essay on the picturesque, as compared with the sublime and the beautiful* (London, J. Robson, 1794).

Raspe, Rudolf Erich, 'Nachricht von einigen niederhessischen Basalten', *Schrifter der Königlicher Societät der Wissenschaften in Göttingen* 1 (1771), p.72-83.

Réimpression de l'Ancien Moniteur, seule histoire authentique et inaltérée de la Révolution française depuis la réunion des Etats-Généraux jusqu'au Consulat (mai 1789-novembre 1799), 32 vols (Paris, 1858-1863).

Reynolds, Frances, *Enquiry concerning the principles of taste, and the origin of our ideas of beauty, etc.* (London, Baker and Galabin, 1785).

Richard, Jérôme, abbé, *Description historique et critique de l'Italie, ou Nouveaux Mémoires sur l'état actuel de son gouvernement, des sciences, des arts, des commerces, de la population, et de l'histoire naturelle*, 6 vols (Paris, Delalain, 1770).

Riedesel, Johann Hermann von, *Reise durch Sicilien und Großgriechenland* (Zurich, Orell, Gessner, Füesslin, 1771).

Roberts, James, 'The Iceland journal of James Roberts', in *Sir Joseph Banks, Iceland and the North Atlantic 1772-1820: journals, letters and documents*, ed. Anna Agnarsdóttir (London, 2016), p.115-40.

Robespierre, Maximilien, *Œuvres de Robespierre*, 10 vols (Paris, 1912-1967).

Roche, Edouard, *Recherches sur les offuscations du soleil et les météores cosmiques* (Paris, 1868).

Roland, Marie-Jeanne, *Mémoires de Madame Roland* (Paris, 1864).

Roland de La Platière, Jean-Marie, *Lettres écrites de Suisse, d'Italie, de Sicile et de Malthe*, 6 vols (Amsterdam, n.n., 1780).

Rousseau, Jean-Jacques, *Discours sur l'origine et les fondemens de l'inégalité parmi les hommes*, in *Collection complète des œuvres*, 17 vols (Geneva, Société typographique de Genève; Barde, Manget & Cie, 1780-1789), vol.1, p.1-176.

–, *Du contrat social, ou Principes du droit politique*, in *Collection complète des œuvres*, 17 vols (Geneva, Société typographique de Genève; Barde, Manget & Cie, 1780-1789), vol.1, p.187-360.

Ruault, Nicolas, *Gazette d'un Parisien sous la Révolution: lettres à son frère, 1783-1796* (Paris, 1976).

Sade, Donatien Alphonse François de, *Histoire de Juliette, ou les Prospérités du vice*, 2 vols (Paris, 1987).

–, *La Nouvelle Justine, ou les Malheurs de la vertu*, 2 vols (Paris, 1987).

–, *Voyage d'Italie* (Paris, 1995).

Saint-Just, Antoine-Louis de, 'Discours du 9 thermidor An II', in *Œuvres complètes* (Paris, 2004), p.769-85.

–, *Organt*, in *Œuvres complètes* (Paris, 2004), p.119-300.

–, 'Rapport au nom du Comité de salut public et du Comité de sûreté générale sur la conjuration ourdie depuis plusieurs années', in *Œuvres complètes* (Paris, 2004), p.706-37.

Saint-Non, Jean-Claude Richard de, *Voyage pittoresque, ou Description des royaumes de Naples et de Sicile*, 5 vols (Paris, Clousier, 1781-1786).

Saussure, Horace-Bénédict de, *Voyages dans les Alpes*, 4 vols (Neuchâtel, Samuel Fauche, 1779-1796).

Scheuchzer, Johann Jakob, *Physica sacra*, 4 vols (Augsburg and Ulm, Christian Ulrich Wagner, 1731-1735).

–, *Physique sacrée, ou Histoire naturelle de la Bible*, 8 vols (Amsterdam, Pierre Schenk; Pierre Mortier, 1732-1737).

Scotti, Emanuele, *Della eruzione del Vesuvio, accaduta il dì 15 giugno 1794* (Naples, Raffaele Lanciano, 1794).

Sherlock, Martin, *Letters from an English traveller* (London, J. Nichols *et al.*, 1780).

Smith, Sir James Edward, *A Sketch of a tour on the continent, in the years 1786 and 1787*, 3 vols (London, n.n., 1793).

Spallanzani, Lazzaro, *Viaggi alle Due Sicilie e in alcune parti dell'Appennini*, 6 vols (Pavia, Baldassare Comini, 1792-1797).

Stanley, Sir John Thomas, *The Journals of the Stanley expedition to the Faroe Islands and Iceland in 1789*, ed. John West, 3 vols (Torshavn, 1970-1976).

Star 1845 (Friday 18 July 1794).

Starke, Mariana, *Letters from Italy, between the years 1792 and 1798, containing a view on the revolutions in that country*, 2 vols (London, 1800).

Steingrímsson, Jón, *Fires of the Earth, the Laki eruption 1783-1784*, translated by Keneva Kunz (Reykjavík, 1998).

–, *A Very present help in trouble: the autobiography of the Fire-Priest*, translated by Michael Fell (New York, 2002).

Sun 560 (Tuesday 15 July 1794).

Swinburne, Henry, *Travels in the Two Sicilies, by Henry Swinburne esq. in the years 1777, 1778, 1779, and 1780*, 2nd edn, 4 vols (London, n.n., 1790).

Tedeschi, Tomaso, *Breve raguaglio degl'incendi di Mongibello avvenuti in quest'anno 1669* (Reggio di Calabria, Domenico Ferro, 1669).

Thomas, Antoine-Léonard, *Essai sur le caractère, les mœurs et l'esprit des femmes dans les différens siècles* (Paris, Moutard, 1772).

Thomson, James, *The Castle of indolence. an allegorical poem, written in imitation of Spenser* (London, n.n., 1748).

Torcia, Michele, 'Lettre sur le tremblement de terre de Calabre', *Le Journal des sçavans* (December 1783), p.839-41.

Torrubia, José, *Disertación histórica política, y en mucha parte geográfica de la islas Filipinas* (Madrid, n.n., 1753).

True Briton (1793) 491 (Friday 25 July 1794).

Valcour, Aristide, *Les Petits Montagnards, opéra-bouffon* (Paris, Cailleau, 1794).

Veryard, Ellis, *An Account of divers choice remarks, as well geographical, as historical, political, mathematical, physical, and moral, taken in a journey through the Low-Countries, France, Italy, and part of Spain, with the Isles of Sicily and Malta* (London, S. Smith and B. Walford, 1701).

Vigée-Lebrun, Louise-Elisabeth, *Souvenirs*, 3 vols (Paris, 1835-1837).

Voltaire, *Candide, ou l'Optimisme*, ed. René Pomeau, in *The Complete works of Voltaire*, vol.48 (Oxford, 1980).

–, *Correspondence and related documents*, ed. Th. Besterman, in *The Complete works of Voltaire*, vol.85-135 (Oxford, 1968-1977).

von Troil, Uno, *Letters on Iceland: containing observations on the civil, literary, ecclesiastical, and natural history, antiquities, volcanos, basaltes, hot springs, customs, dress, manners of the inhabitants, &c., &c.* (London, W. Richardson, 1780).

The Vulcano's, or Burning and fire-vomiting mountains, famous in the world (London, J. Darby, 1669).

Walpole, Horace, *Horace Walpole's correspondence*, ed. W. S. Lewis *et al.*, 48 vols (London and Oxford, 1937-1983).

Watkins, Thomas, *Travels through Switzerland, Italy, Sicily, the Greek islands to Constantinople, through part of Greece, Ragusa and the Dalmatian isles in the years 1787, 1788, 1789*, 2 vols (London, J. Owen, 1794).

Werner, Abraham Gottlob, *Kurze Klassifikation und Beschreibung der verschiedenen Gebirgsarten* (Dresden, Walther, 1787).

White, Gilbert, *The Natural history and antiquities of Selbourne, in the county of Southampton: with*

engravings, and an appendix (London, T. Bensley, 1789).
Williams, Helen Maria, *Letters containing a sketch of the politics of France, from the thirty-first of May 1793, till the twenty-eight of July 1794, and of the scenes which have passed in the prisons of Paris*, 2 vols (London, G. G. and J. Robinson, 1795).
Winckelmann, Johann Joachim, *Briefe*, 4 vols (Berlin, 1952-1957).
–, *Sendschreiben von den Herculanischen Entdeckungen* (Dresden, n.n., 1762).

Works written after 1850

Abramovici, Jean-Christophe, 'Sade et l'irruption du désir', in *Le Vésuve en éruption: savoirs, représentations, pratiques*, ed. Emilie Beck Saiello and Dominique Bertrand (Clermont-Ferrand, 2013), p.171-77.
Agnarsdóttir, Anna, 'Iceland in the eighteenth century: an island outpost of Europe?', *Sjuttonhundratal* (2013), p.11-38.
–, 'Iceland under British protection during the Napoleonic wars', in *Scandinavia in the age of revolution: Nordic political cultures, 1740-1820*, ed. Pasi Ihalainen *et al.* (Farnham, 2011), p.255-66.
–, 'Introduction: Banks and Iceland', in *Sir Joseph Banks, Iceland and the North Atlantic 1772-1820: journals, letters and documents*, ed. Anna Agnarsdóttir (London, 2016), p.1-34.
Albigès, Luce-Marie, 'La fête de l'Etre Suprême, 20 prairial An II (8 juin 1794)', http://www.histoire-image.org/etudes/fete-etre-supreme-20-prairial-ii-8-juin-1794 (last accessed 14 December 2018).
Alcantara, Jean-Pascal, 'De neige et de lave mêlées: le paysage volcanique d'après Horace-Bénédict de Saussure', in *L'Invention du paysage volcanique*, ed. Dominique Bertrand (Clermont-Ferrand, 2004), p.83-102.
Ashburn Miller, Mary, *A Natural history of revolution: violence and nature in the French revolutionary imagination 1789-1794* (Ithaca, NY, and London, 2011).
Aulard, François-Alphonse, *Les Orateurs de la Législative et de la Convention*, 2 vols (Paris, 1885-1886).
–, *Les Orateurs de l'Assemblée constituante* (Paris, 1882).

Bachelard, Gaston, *La Psychanalyse du feu* (Paris, 1949).
Baere, Benoît de, 'Natural catastrophe in Buffon's *Histoire naturelle*: earth, science, aesthetics, anthropology', in *Histoires de la Terre: earth sciences and French culture 1740-1940*, ed. Louise Lyle and David McCallam (Amsterdam, 2008), p.19-35.
Barnett, Lydia, 'The theology of climate change: sin as agency in the Enlightenment's Anthropocene', *Environmental history* 20 (2015), p.217-37.
Battersby, Christine, *The Sublime, terror and human difference* (London, 2007).

–, 'Terror, terrorism and the sublime: rethinking the sublime after 1789 and 2001', *Postcolonial studies* 6:1 (2003), p.67-89.

Beck Saiello, Emilie, *Pierre Jacques Volaire 1729-1799, dit le chevalier Volaire* (Paris, 2010).

–, and Brice Gruet, 'De la girandole du château Saint-Ange au Vésuve en éruption: voyage d'une métaphore visuelle', in *Le Vésuve en éruption: savoirs, représentations, pratiques*, ed. Emilie Beck Saiello and Dominique Bertrand (Clermont-Ferrand, 2013), p.151-67.

Bell, C. F., and Thomas Girtin, 'The drawings and sketches of John Robert Cozens: a catalogue with a historical introduction', *The Volume of the Walpole Society* 23 (1934-1935), p.i-xi, 1-87; including 'Some additions and notes to the twenty-third volume of the Walpole Society', p.1-12.

Bermingham, Ann, 'Landscape-o-rama: the exhibition landscape at Somerset House and the rise of popular landscape entertainments', in *Art on the line: the Royal Academy exhibitions at Somerset House, 1780-1836* (New Haven, CT, 2001), p.127-43.

Björnsson, Helgi, *The Glaciers of Iceland: a historical, cultural and scientific overview* (Paris and Amsterdam, 2017).

Black, Jeremy, *Eighteenth-century Europe*, 2nd edn (Basingstoke, 1999).

Bodei, Remo, *Paesaggi sublimi: gli uomini davanti alla natura selvaggia* (Milan, 2008).

Bonnet, Jean-Claude, *Naissance du Panthéon: essai sur le culte des grands hommes* (Paris, 1998).

Borgongino, Michele, and Grete Stefani, 'Intorno alla data dell'eruzione del 79 d.C', *Rivista di studi pompeiani* 12-13 (2001-2002), p.177-215.

Bowen, Margarita, *Empiricism and geographical thought: from Francis Bacon to Alexander von Humboldt* (Cambridge, 2009).

Brewer, John, 'The kingdom of Vesuvius', https://brewersblog.org/page/1/ (last accessed 6 December 2018).

–, 'Sublime tourism, enlightened science and counter-revolution: three versions of Vesuvius and Pompeii in the age of Romanticism', research presentation given at the University of Durham, 25 February 2016, https://www.dur.ac.uk/resources/ias/audio/Brewer.mp3 (last accessed 3 December 2018).

Broc, Numa, *Les Montagnes au siècle des Lumières, perception et représentation* (Paris, 1991).

Budd, Malcolm, 'Delight in the natural world: Kant on the æsthetic appreciation of nature, Part III: the sublime in nature', *British journal of æsthetics* 38:3 (July 1998), p.233-50.

Calaresu, Melissa, 'From the street to stereotype: urban space, travel and the picturesque in late eighteenth-century Naples', *Italian studies* 62:2 (autumn 2007), p.189-203.

–, 'Making and eating ice cream in Naples: rethinking consumption and sociability in the eighteenth century', *Past and present* 220 (August 2013), p.35-78.

–, and Helen Hills (ed.), *New approaches to Naples c.1500-c.1800: the power of place* (Abingdon, 2016).

Candela, Andrea, 'Recupero, Giuseppe', in *Dizionario biografico degli Italiani*, vol.86 (2016), http://www.treccani.it/enciclopedia/giuseppe-recupero_(Dizionario-Biografico)/ (last accessed 7 December 2018).

Capasso, Luigi, *I Fuggiaschi di Ercolano: paleobiologia delle vittime dell'eruzione vesuviana del 79 d.C.* (Rome, 2001).

Carozzi, Albert V., 'De Maillet's *Telliamed* (1748): an ultra-Neptunian theory of the Earth', in *Toward a history of geology*, ed. Cecil J. Schneer (Cambridge, MA, 1969), p.80-99.

–, 'Horace-Bénédict de Saussure pionnier en 1784 du concept de "refoulements horizontaux en sens contraires" en géologie structurale', *Travaux du Comité français d'histoire de la géologie* 3:11 (1977), p.89-112.

Carter, Harold B., *Sir Joseph Banks 1743-1820* (London, 1988).

Casapullo, Rosa, 'Note sull'italiano della vulcanologia fra Seicento e Settecento', in *Napoli e il Gigante: il Vesuvio tra imagine, scrittura e memoria*, ed. Rosa Casapullo and Lorenza Gianfrancesco (Soveria Mannelli, 2014), p.13-53.

Cassirer, Ernst, *The Philosophy of Enlightenment* (Princeton, NJ, 1951).

Castle, Terry, *The Female thermometer: eighteenth-century culture and the invention of the uncanny* (Oxford, 1995).

Castonguay-Bélanger, Joël, 'Le sort de Galilée: *Paul et Virginie* et la théorie des marées de Bernardin de Saint-Pierre', *Eighteenth-century fiction* 20 (2007-2008), p.177-96.

Castro, Americo, 'Rousseau and the Alps', in *An Idea of history: selected essays of Americo Castro*, ed. Stephen Gilman and Edmund L. King (Columbus, OH, 1977), p.271-76.

Chard, Chloe, *Pleasure and guilt on the Grand Tour: travel writing and imaginative geography 1600-1830* (Manchester, 1999).

Chester, David K., and Angus M. Duncan, 'Geomythology, theodicy, and the continuing relevance of religious worldviews on responses to volcanic eruptions', in *Living under the shadow: the cultural impacts of volcanic eruptions*, ed. John Grattan and Robin Torrence (Walnut Creek, CA, 2007), p.203-24.

Chevallier, Elisabeth, 'La finalité du Vésuve: force du mal ou instrument de la providence selon les étrangers venus à Naples au XVIIIe siècle', in *Colloque histoire et historiographie: Clio*, ed. Raymond Chevallier (Paris, 1980), p.349-69.

Ciancio, Luca, *Autopsie della Terra: illuminismo e geologia in Alberto Fortis (1741-1803)* (Florence, 1995).

Cocco, Sean, *Watching Vesuvius: a history of science and culture in early modern Italy* (Chicago, IL, 2013).

Combeau, Yvan, 'De Bourbon à La Réunion, l'histoire d'une île (du XVIIe au XXe siècle)', *Hermès* 32-33 (2002), p.91-99.

Constantine, David, *Fields of fire: a life of Sir William Hamilton* (London, 2001).

–, 'Winckelmann and Sir William Hamilton', *Oxford German studies* 22 (1993), p.55-83.

Cooper, John, 'James Gillray and the French Revolution', *RSA journal* 137 (September 1989), p.646-51.

Corsi, Pietro, *The Age of Lamarck: evolutionary theories in France 1790-1830* (Berkeley, CA, 1988).

Courtois, Jean-Patrice, 'Le climat chez Montesquieu et Rousseau', in *L'Evénement climatique et ses représentations (XVIIe-XIXe siècle)*, ed. Emmanuel Le Roy Ladurie *et al.* (Paris, 2007), p.157-80.

Crutzen, Paul J., and Eugene F. Stoermer, 'The "Anthropocene"', *IGBP newsletter* 41 (May 2000), p.17-18.

Csergo, Julia, 'The picnic in nineteenth-century France: a social event involving food – both a necessity and a form of entertainment', in *Eating out in Europe: picnics, gourmet dining and snacks since the late eighteenth century*, ed. Marc Jacobs and Peter Scholliers (Oxford, 2003), p.139-59.

Curry, Andrew, 'That time a German prince built an artificial volcano', *Smithsonian.com* (29 August 2012), https://www.smithsonianmag.com/history/that-time-a-german-prince-built-an-artificial-volcano-19001853/ (last accessed 10 December 2018).

D'Amore, Manuela, *The Royal Society and the discovery of the Two Sicilies: southern routes in the Grand Tour* (Basingstoke, 2017).

Darley, Gillian, *Vesuvius: the most famous volcano in the world* (London, 2012).

D'Arrigo, Rosanne, *et al.*, 'The anomalous winter of 1783-1784: was the Laki eruption or an analog of the 2009-2010 winter to blame?', *Geophysical research letters* 38 (2011), L05706, doi:10.1029/2011GL046696 (last accessed 17 December 2018).

David, Elizabeth, *Harvest of the cold months: the social history of ice and ices* (London, 1994).

De Carolis, Ernesto, and Giovanni Patricelli, *Vesuvio 79 d.C. La distruzione di Pompei ed Ercolano* (Rome, 2003).

De Kesel, Marc, '"Wie der Gott und Mensch sich paart": Hölderlin's theory of the sublime as read in his "Remarks on tragedy"', in *Histories of the sublime*, ed. Bart Vandenabeele (Brussels, 2005), p.123-33.

de Vries, Jan, *The Industrious revolution: consumer behavior and the household economy, 1650 to the present* (Cambridge, 2008).

Dommanget, Maurice, *Sylvain Maréchal: l'égalitaire, 'l'homme sans Dieu', 1750-1803 – vie et œuvre de l'auteur du Manifeste des égaux* (Paris, 1950).

Doran, Robert, *The Theory of the sublime from Longinus to Kant* (Cambridge, 2015).

Dubost, Jean-Pierre, 'The sublime adverse and its Sadian reverse: Kant, Sade and Schiller', *Eighteenth-century studies* 28:1 (1994), p.83-93.

Duffy, Cian, *The Landscapes of the sublime 1700-1830: classic ground* (Basingstoke, 2013).

Dupuy, Pascal, and Caroline Bourgeois, 'Les représentations de la République française à travers les gravures satiriques (1792-1799) et la presse quotidienne anglaise (1792-1793)', *Annales historiques de la Révolution française* 289 (1992), p.367-81.

Ecker, Hans-Peter, 'Vergewisserung subjektiver Identität, Spurensuche im Kollektiven Gedächtnis, Horrortrip in die Wildnis: über kulturelle Programme literarischer Harzreisen', *Zeitschrift für Germanistik* 7:2 (1997), p.262-77.

Electronic Enlightenment scholarly edition of correspondence.

Favier, René, 'Les éruptions du Vésuve dans la littérature populaire de la seconde moitié du XVIIIe siècle', in *Le Vésuve en éruption: savoirs, représentations, pratiques*, ed. Emilie Beck Saiello and Dominique Bertrand (Clermont-Ferrand, 2013), p.99-109.

Findlen, Paula (ed.), *Athanasius Kircher: the last man who knew everything* (New York, 2004).

Foucault, Michel, *Les Mots et les choses: une archéologie des sciences humaines* (Paris, 1966).

–, *Surveiller et punir* (Paris, 1975).

Fowler, C. M. R., *The Solid Earth: an introduction to global geophysics*, 2nd edn (Cambridge, 2005).

France, Peter, 'Speakers and audience: the first days of the Convention', in *Language and rhetoric of the Revolution*, ed. John Renwick (Edinburgh, 1990), p.50-74.

Freud, Sigmund, 'The "uncanny"' (1919), in *The Standard edition of the complete psychological works of Sigmund Freud*, 24 vols (London, 1953-1974), vol.17, p.217-52.

Fried, Michael, *Absorption and theatricality: painting and beholder in the age of Diderot* (London, 1980).

Fumaroli, Marc (ed.), *Histoire de la rhétorique dans l'Europe moderne, 1450-1950* (Paris, 1999).

Furet, François, *Penser la Révolution française* (Paris, 1978).

Gardner Coates, Victoria C., and Jon L. Seydl (ed.), *Antiquity recovered: the legacy of Pompeii and Herculaneum* (Los Angeles, CA, 2007).

Garnier, Emmanuel, 'Les brouillards du Laki en 1783: volcanisme et crise sanitaire en Europe', *Bulletin de l'Académie nationale de médecine* 195:4-5 (2011), p.1043-55.

Gascoigne, John, *Joseph Banks and the English Enlightenment: useful knowledge and polite culture* (Cambridge, 2003).

Gauchet, Marcel, *La Révolution des droits de l'homme* (Paris, 1989).

Gaudreau, Nicolas, 'La gloire du peintre et les errements de l'Académie: des pistes pour l'étude des marines de Claude-Joseph Vernet au salon', *RACAR: revue d'art canadienne / Canadian art review* 27:1-2 (2000), p.74-86.

Gilby, Emma, *Sublime worlds: early modern French literature* (London, 2006).

Godwin, Joscelyn, *Athanasius Kircher: a Renaissance man and the quest for lost knowledge* (London, 1979).

Goodden, Angelica, *Miss Angel: the art and world of Angelica Kaufman, eighteenth-century icon* (London, 2006).

Grant, Michael, *Cities of Vesuvius: Pompeii and Herculaneum* (London, 2005).

Grattan, John, and Mark Brayshay, 'An amazing and portentous summer: environmental and social responses in Britain to the 1783 eruption of an Iceland volcano', *The Geographical journal* 161:2 (1995), p.125-34.

Grattan, John, Sabina Michnowicz and Robin Rabartin, 'The long shadow: understanding the influence of the Laki fissure eruption on human mortality in Europe', in *Living under the shadow: the cultural impacts of volcanic eruptions*, ed. John Grattan and Robin Torrence (Walnut Creek, CA, 2007), p.153-74.

Grattan, John, *et al.*, 'Volcanic air pollution and mortality in France 1783-1784', *Comptes rendus geoscience* 337 (2005), p.641-51.

Greene, Mott T., *Geology in the nineteenth century: changing views of a changing world* (Ithaca, NY, and London, 1982).

Grell, Chantal, *Herculanum et Pompéi dans les récits des voyageurs français du XVIII*[e] *siècle* (Naples, 1982).

–, 'Regarder, observer et interroger le Vésuve: le volcan comme "objet d'étude" au XVIIIe siècle', in *Le Vésuve en éruption: savoirs, représentations, pratiques*, ed. Emilie Beck Saiello and Dominique Bertrand (Clermont-Ferrand, 2013), p.51-81.

Grignon, Claude, 'Commensality and social morphology: an essay in typology', in *Food, drink and identity: cooking, eating and drinking in Europe since the Middle Ages*, ed. Peter Scholliers (Oxford, 2001), p.23-33.

Grove, Richard, *Green imperialism: colonial expansion, tropical island Edens and the origins of environmentalism, 1600-1860* (Cambridge, 1996).

Guerra, Corinna, 'If you don't have a good laboratory, find a good volcano: Mount Vesuvius as a natural chemical laboratory in eighteenth-century Italy', *Ambix* 62:3 (August 2015), p.245-65.

Guilhem, Armand, '*Le Spectacle de la nature* ou l'esthétique de la Révélation', *Dix-huitième siècle* 45:1 (2013), p.329-45.

Guillaume, Eric, *et al.*, 'Considerations on combustion and fire behaviour of materials: a change of mind during the 18th century', *Journal of fire studies* 34:1 (2016), p.69-84.

Hales, Shelley, and Joanna Paul (ed.), *Pompeii in the public imagination, from rediscovery to today* (Oxford, 2011).

Hamblyn, Richard, *The Invention of clouds: how an amateur meteorologist forged the language of the skies* (London, 2001).

Hamiche, Daniel, *Le Théâtre et la Révolution* (Paris, 1973).

Hamilton, James, 'Les volcans dans l'art et la littérature britannique au XIXe siècle: un bref survol', in *Le Vésuve en éruption: savoirs, représentations, pratiques*, ed. Emilie Beck Saiello and Dominique Bertrand (Clermont-Ferrand, 2013), p.129-39.

Heilbron, John Lewis, *Electricity in the 17th and 18th centuries: a study of early modern physics* (Berkeley, CA, 1979).

Heringman, Noah, *Sciences of antiquity: romantic antiquarianism, natural history, and knowledge work* (Oxford, 2013).

–, 'The style of natural catastrophes', *The Huntingdon Library quarterly* 66:1-2 (2003), p.97-133.

Hinnaut, Charles, '"The late unfortunate regicide in France": Burke and the political sublime', *1650-1850: ideas, aesthetics, and inquiries in the early modern era* 2 (1996), p.111-36.

Hitt, Christopher, 'Towards an ecological sublime', *New literary history* 30:3 (summer 1999), p.603-23.

Holmes, Richard, *The Age of wonder: how the Romantic generation discovered the beauty and terror of science* (London, 2008).

Hooykaas, Reijer, *Catastrophism in geology, its scientific character in relation to actualism and uniformitarianism* (Amsterdam, 1970).

Hose, T. A., 'Towards a history of geotourism: definitions, antecedents and the future', in *The History of geoconservation*, ed. C. V. Burek and C. D. Prosser (London, 2008), p.37-60.

Huet, Marie-Hélène, 'Le sacre du printemps: essai sur le sublime et la Terreur', *Modern language notes* 103:4 (1988), p.782-99.

Israel, Jonathan I., *Enlightenment contested: philosophy, modernity and the emancipation of man 1670-1752* (Oxford, 2006).

–, *Radical Enlightenment: philosophy and the making of modernity 1650-1750* (Oxford, 2001).

Jacobelli, Luciana, *Pompei: la costruzione di un mito* (Rome, 2008).

Jenkins, Ian, and Kim Sloan (ed.), *Vases and volcanoes: Sir William Hamilton and his collection* (London, 1996).

Jerram, Dougal, Alwyn Scarth and Jean-Claude Tanguy, *Volcanoes of Europe*, 2nd edn (Edinburgh, 2017).

Kennedy, Barbara A., *Inventing the Earth: ideas on landscape development since 1740* (Oxford, 2006).

Kenny, Neil, *The Uses of curiosity in early modern France and Germany* (Oxford, 2004).

Kerslake, Lawrence, 'Aspects of the epistemology of the sublime in seventeenth- and eighteenth-century French theories', *Phrasis: studies in language and literature* 46 (2005), p.7-29.

Knight, Carlo, 'Sir William Hamilton's *Campi Phlegraei* and the artistic contribution of Peter Fabris', in *Oxford, China and Italy: writings in honour of Sir Harold Acton* (London, 1984), p.192-208.

Krul, Wessel, 'Style and ideology in Edmund Burke's concept of the sublime', *Phrasis: studies in language and literature* 46 (2005), p.31-41.

Lamb, Jonathan, 'Inchoate possession: how Captain Kerguelen claimed an island', *Journal for maritime research* 7:1 (January 2005), p.1-15.

Langdon, Helen, Xavier Salomon and Caterina Volpi, *Salvator Rosa: bandits, wilderness and magic* (London, 2010).

Leask, Nigel, *Curiosity and the aesthetics of travel writing, 1770-1840* (Oxford, 2002).

Lénat, Jean-François, 'A brief history of the observation of the central area of Piton de la Fournaise', in *Active volcanoes of the southwest Indian Ocean: Piton de la Fournaise and Karthala*, ed. Patrick Bachèlery *et al.* (Heidelberg, 2016), p.3-21.

Le Roy Ladurie, Emmanuel, *Histoire humaine et comparée du climat II: disettes et révolutions 1740-1860* (Paris, 2006).

Linton, Marisa, *Choosing Terror: virtue, friendship, and authenticity in the French Revolution* (Oxford, 2013).

Longeot, Denise, 'La genèse de la cristallographie et de la minéralogie scientifique', *Dix-huitième siècle* 3 (1971), p.253-64.

Loussouarn, Sophie, 'Gillray and the French Revolution', *National identities* (June 2015), p.1-17.

McCallam, David, 'From analogies to patterns: images in French and British geological texts (1760-1800)', in *Visualizing the text: from manuscript culture to the age of caricature*, ed. Lauren Beck and Christina Ionescu (Newark, DE, 2017), p.203-27.

–, 'Quelques réflexions sur la figure de la chaîne au XVIIIe siècle: en théorie et en pratique', in *De la théorie à l'action: les savoirs et leur mise en œuvre au siècle des Lumières*, ed. Valérie Kobi (Geneva, 2011), p.123-34.

Madelein, Christoph, and Jürgen Pieters, 'The sublime re-considered: by way of introduction', *Phrasis: studies in language and literature* 46 (2005), p.1-6.

Malm, Mats, 'On the technique of the sublime', *Comparative literature* 52:1 (winter 2000), p.1-10.

Marcil, Yasmine, 'Découvrir, comprendre, ressentir la montagne dans la presse périodique des années 1780', *Compar(a)ison: an international journal of comparative literature* 1-2 (2001), p.145-70.

Mercier-Faivre, Anne-Marie, 'Eruptions dans la presse: l'image du Vésuve dans la presse européenne francophone du XVIIIe siècle', in *Le Vésuve en éruption: savoirs, représentations, pratiques*, ed. Emilie Beck Saiello and Dominique Bertrand (Clermont-Ferrand, 2013), p.83-98.

Messina, Simone, 'Le naturaliste et la catastrophe: Dolomieu en Calabre, 1784', in *L'Invention de la catastrophe au XVIIIe siècle: du châtiment divin au désastre naturel*, ed.

Anne-Marie Mercier-Faivre and Chantal Thomas (Geneva, 2008), p.285-302.

Mikhail, Alan, 'Ottoman Iceland: a climate history', *Environment history* 20 (2015), p.262-84.

Molesworth, Jesse, 'The cosmic sublime: Wright of Derby's *A Philosopher lecturing on the orrery*', *Lumen* 34 (2015), p.109-21.

Moore, Andrew, and Nigel Larkin, *Art at the rockface: the fascination of stone* (London, 2006).

Moore, James, 'History as theoretical reconstruction? Baron d'Hancarville and the exploration of ancient mythology in the eighteenth century', in *Reinventing history: the Enlightenment origins of ancient history*, ed. James Moore, Ian Macgregor Morris and Andrew J. Bayliss (London, 2008), p.137-67.

Nadeau, Martin, 'La politique culturelle de l'An II: les infortunes de la propagande révolutionnaire au théâtre', *Annales historiques de la Révolution française* 327 (January-March 2002), p.57-74.

Nazzaro, Antonio, 'Implicazioni di una ermeneutica delle fonti vesuviane sull'eruzione del 1631: forma del vulcano e risposta al rischio', in *Napoli e il Gigante: il Vesuvio tra imagine, scrittura e memoria*, ed. Rosa Casapullo and Lorenza Gianfrancesco (Soveria Mannelli, 2014), p.137-78.

Neall, Vincent E., 'Volcanic soils', in *Encyclopedia of land use, land cover and soil sciences*, vol.7, ed. Willy H. Verheye (Oxford, 2006), p.23-45.

Nicolson, Benedict, *Joseph Wright of Derby: painter of light*, 2 vols (London, 1968).

Nicolson, Marjorie Hope, *Mountain gloom and mountain glory: the development of the aesthetics of the infinite* (Seattle, WA, 1997).

O'Dea, Michael, 'Le mot "catastrophe"', in *L'Invention de la catastrophe au XVIIIe siècle: du châtiment divin au désastre naturel*, ed. Anne-Marie Mercier-Faivre and Chantal Thomas (Geneva, 2008), p.35-48.

Ogilvie, Sheilagh C., and Markus Cerman (ed.), *European proto-industrialization* (Cambridge, 1996).

Oldroyd, David R., *Thinking about the Earth: a history of ideas in geology* (London, 1996).

Oppenheimer, Clive, *Eruptions that shook the world* (Cambridge, 2011).

Oslund, Karen, *Iceland imagined: nature, culture, and storytelling in the North Atlantic* (Seattle, 2011).

Oxford English dictionary online (3rd edn, March 2013), http://www.oed.com.sheffield.idm.oclc.org/ (last accessed 3 December 2018).

Ozouf, Mona, *Festivals and the French Revolution*, translated by Alan Sheridan (Cambridge, MA, 1988).

Palanciuc, Luiza, 'Figures de l'éloignement et paysages volcaniques dans la peinture européenne', in *L'Invention du paysage volcanique*, ed. Dominique Bertrand (Clermont-Ferrand, 2004), p.187-98.

Parker, Kevin, 'Winckelmann, historical difference, and the

problem of the boy', *Eighteenth-century studies* 25:4 (1992), p.523-44.

Perović, Sanja, *The Calendar in Revolutionary France: perceptions of time in literature, culture, politics* (Cambridge, 2012).

Pierce, Gillian B., *Scapeland: writing the landscape from Diderot's Salons to the postmodern museum* (Amsterdam, 2012).

Pinto, Manuel S., and Filomena Amador, 'Discussing the age of the Earth in 1779 in Portugal', in *Geology and religion: a history of harmony and hostility*, ed. M. Kölbl-Ebert (London, 2009), p.83-87.

Placanica, Augusto, *Il Filosofo e la catastrofe: un terremoto del Settecento* (Turin, 1985).

Porter, Roy, *The Making of geology: earth science in Britain 1660-1815* (Cambridge, 1977).

Principato, Aurelio, 'Comment restituer l'action oratoire de la Révolution?', in *Une Expérience rhétorique: l'éloquence de la Révolution*, ed. Eric Négrel and Jean-Paul Sermain (Oxford, 2002), p.19-33.

–, 'L'éloquence révolutionnaire: idéologie et légende', in *Histoire de la rhétorique dans l'Europe moderne, 1450-1950*, ed. Marc Fumaroli (Paris, 1999), p.1019-37.

Pyle, David M., *Volcanoes: encounters through the ages* (Oxford, 2017).

Quenet, Grégory, *Qu'est-ce que l'histoire environnementale?* (Seyssel, 2014).

–, *Les Tremblements de terre aux XVII*[e] *et XVIII*[e] *siècles: la naissance d'un risque* (Seyssel, 2005).

–, 'When geology encounters a real catastrophe: from theoretical earthquakes to the Lisbon disaster', in *Histoires de la Terre: earth sciences and French culture 1740-1940*, ed. Louise Lyle and David McCallam (Amsterdam, 2008), p.37-56.

Quinzio, Jeri, *Of sugar and snow: a history of ice cream making* (Berkeley, CA, 2009).

Racault, Jean-Michel, 'Géologie, vulcanologie et imaginaire chez Bernardin de Saint-Pierre', *Revue italienne d'études françaises* 1 (2011), p.37-54.

Ramage, Nancy H., 'Goods, graves, and scholars: 18th-century archaeologists in Britain and Italy', *American journal of archaeology* 96 (1992), p.653-61.

Rappaport, Rhoda, 'The geological atlas of Guettard, Lavoisier and Monnet: conflicting views of the nature of geology', in *Toward a history of geology*, ed. Cecil J. Schneer (Cambridge, MA, 1969), p.272-87.

–, *When geologists were historians, 1665-1750* (Ithaca, NY, and London, 1997).

Reichardt, Rolf, and Hubertus Kohle, *Visualizing the Revolution: politics and the pictorial arts in late eighteenth-century France* (London, 2008).

Rigby, Kate, *Dancing with disaster: environmental histories, narratives, and ethics for perilous times* (Charlottesville, VA, 2015).

Roberts, Paul, *Life and death in Pompeii and Herculaneum* (London, 2013).

Rodríguez Fernández, Laura, 'Il barocco spagnolo di fronte al Vesuvio: Juan de Quiñones, *El monte Vesuvio, aora la montaña de Soma* (Juan González, Madrid, 1632)', in *Napoli e il Gigante: il Vesuvio tra imagine, scrittura e memoria*, ed. Rosa Casapullo and Lorenza Gianfrancesco (Soveria Mannelli, 2014), p.111-36.

Roos, Anna Marie, 'Martin Lister (1639-1712) and fools' gold', *Ambix* 51:1 (2004), p.23-41.

Rudwick, Martin J. S., *Bursting the limits of time: the reconstruction of geohistory in the age of revolution* (Chicago, IL, 2005).

–, 'Geological travel and theoretical innovation: the role of "liminal" experience', *Social studies in science* 26:1 (1996), p.143-59.

–, *The New science of geology: studies in the earth sciences in the age of revolution* (Aldershot, 2004).

Saint Girons, Baldine, *Fiat lux: une philosophie du sublime* (Paris, 1993).

Scarth, Alwyn, *Vesuvius: a biography* (Princeton, NJ, 2009).

–, *Vulcan's fury: man against the volcano* (New Haven, CT, 2001).

Schama, Simon, *Citizens: a chronicle of the French Revolution* (London, 1989).

Schmidt, Anja, *et al.*, 'Climatic impact of the long-lasting 1783 Laki eruption: inapplicability of mass-independent sulfur isotopic composition measurements', *Journal of geophysical research* 117 (2012), D23116, doi:10.1029/2012JD018414 (last accessed 17 December 2018).

Schulz, Karsten A., 'Decolonizing the Anthropocene: the mytho-politics of human mastery', in *Critical epistemologies of global politics*, ed. Marc Woons and Sebastian Weier (Bristol, 2017), p.46-62.

Shaw, Philip, *The Sublime* (Abingdon, 2006).

Smith, Denis Mack, *A History of Sicily: modern Sicily, after 1713*, 2 vols (London, 1968).

Smyth, Jonathan, *Robespierre and the Festival of the Supreme Being: the search for a republican morality* (Manchester, 2016).

Sontag, Susan, *The Volcano lover: a romance* (London, 1993).

Steffan, Will, *et al.*, 'The Anthropocene: conceptual and historical perspectives', *Philosophical transactions of the Royal Society* 369 (2011), p.842-67.

Stothers, Richard B., 'The great dry fog of 1783', *Climate change* 32:1 (1996), p.79-89.

–, 'Volcanic dry fogs, climate cooling, and plague pandemics in Europe and the Middle East', *Climatic change* 42 (1999), p.713-23.

Taub, Liba, *Ætna and the moon: explaining nature in ancient Greece and Rome* (Corvallis, OR, 2008).

Taylor, K. L., 'Geological travellers in Auvergne, 1751-1800', in *Four centuries of geological travel: the search for knowledge on foot, bicycle, sledge and camel*, ed. Patrick N. Wyse Jackson (London, 2007), p.73-96.

Thacker, Christopher, 'The volcano: culmination of the landscape garden', *Eighteenth-century life* 8 (January 1983), p.74-83.

Thagard, Paul, *Conceptual revolutions* (Princeton, NJ, 1992).
Thordarson, Thorvaldur, and Stephen Self, 'Atmospheric and environmental effects of the 1783-1784 Laki eruption: a review and reassessment', *Journal of geophysical research: atmospheres* 108 (2003), p.1-29.
Thordarson, Thorvaldur, *et al.*, 'The 1783-1785 A.D. Laki–Grimsvötn eruptions II: appraisal based on comtemporary accounts', *Jökull* 53 (2003), p.11-48.
Thüsen, Joachim von der, 'Juliette at the volcano: reflections on the Sadean journey', *Studies in eighteenth-century culture* 22 (1992), p.87-103.
–, *Schönheit und Schrecken der Vulkane: zur Kulturgeschichte des Vulkanismus* (Darmstadt, 2008).
Todorov, Tzvetan, *L'Esprit des Lumières* (Paris, 2006).

Udías, Augustín, 'Earthquakes as God's punishment in 17th- and 18th-century Spain', in *Geology and religion: a history of harmony and hostility*, ed. M. Kölbl-Ebert (London, 2009), p.41-48.

Vacari, Ezio, *Giovanni Arduino (1714-1795): il contributo di uno scienziato veneto al dibattito settecentesco sulle scienze della Terra* (Florence, 1993).
–, 'Spallanzani e le scienze geologiche del Settecento: un percorso interpretativo tra carteggi e diari di viaggio', in *La Sfida della modernità: atti del convegno di studi nel bicentenario della morte di Lazzaro Spallanzani*, ed. Walter Bernardi and Marta Stefani (Florence, 2000), p.293-317.
Vaclavik, Kiera, *Uncharted depths: descent narratives in English and French children's literature* (Oxford, 2010).
Verne, Jules, *Voyage au centre de la Terre* (Paris, 1864).
Viallon, Marie, 'Le *buon gigante*', in *Le Vésuve en éruption: savoirs, représentations, pratiques*, ed. Emilie Beck Saiello and Dominique Bertrand (Clermont-Ferrand, 2013), p.41-50.

Wallace-Hadrill, Andrew, *Herculaneum: past and future* (London, 2012).
Wawn, Andrew, 'John Thomas Stanley and Iceland: the sense and sensibility of an eighteenth-century explorer', *Scandinavian studies* 53:1 (winter 1981), p.52-76.

Young, Davis A., *Mind over magma: the story of igneous petrology* (Princeton, NJ, 2003).

Zanetti, François, *L'Electricité médicale dans la France des Lumières* (Oxford, 2017).
Zapf, Hubert, 'Cultural ecology, the environmental humanities, and the transdisciplinary knowledge of literature', in *Environmental humanities: voices from the Anthropocene*, ed. Serpil Oppermann and Serenella Iovino (London, 2017), p.61-79.

Index